Upgrade A-Level CHEMISTRY

Ted Lister and Janet Renshaw

Trinity School, Leamington Spa

Stanley Thornes (Publishers) Ltd

First published in 1996 by:
Stanley Thornes (Publishers) Ltd
Ellenborough House
Wellington Street
CHELTENHAM GL50 1YW
England

A catalogue record for this book is available from the British Library.

ISBN 0–7487–2575–X

96 97 98 99 00 / 10 9 8 7 6 5 4 3 2 1

Typeset by Tech-Set Ltd, Tyne & Wear
Printed and bound in Great Britain at Scotprint, Musselburgh

ACKNOWLEDGEMENTS

The authors and publisher would like to thank the following examination boards for permission to reproduce
questions from examination papers.
Northern Examinations and Assessment Board (N)
Oxford and Cambridge Schools Examination Board (O & C)
University of Cambridge Local Examinations Syndicate (C)
University of London Examinations and Assessment Council (L)
University of Oxford Delegacy of Local Examinations (O)
Abbreviations used for the above in the text are shown in brackets.
The examination boards accept no responsibility whatsoever for the accuracy or method of working in the
answers given. These are solely the responsibility of the authors.
The authors would also like to thank Drs Alan Cox and George Raper of Warwick University for useful
discussions, Margaret O'Gorman for her encouragement and John Hepburn for his helpful suggestions and
meticulous editing.

CONTENTS

INTRODUCTION

The aim of this book is, as the title suggests, to enable you to improve the grade you will get at A-level chemistry. You can do this in two ways; by knowing and understanding more chemistry, and by becoming better at doing chemistry exams. This book aims to help in both areas. It covers core topics and some of the more popular options. However, even in the optional topics we have chosen subject matter that relates to mainstream chemistry, so it will be worth tackling even if you are not doing that particular option.

The way this book will help to upgrade your chemistry is by going over a number of A-level questions on a variety of topics and working through the answers. We have talked around the answers putting the chemistry in context, pointing out common mistakes and pitfalls and also what examiners are looking for. As far as is possible in a book, we have tried to do this as though we were sitting down with you as a student, discussing issues which arise from each question.

Of course what you need to know is just what to write on the exam paper to get full marks, so we draw a distinction between this and our commentary by printing the answers in **bold**.

There are tips on exam technique, common mistakes to avoid and 'by the way' sections which tell you about some chemistry which may not be needed for the answer to this question but might just turn up in the paper that you sit. We have also included sections entitled 'Why does it matter?' which try to show you that chemistry is not just about passing exams but is also important in everyday life.

Taking exams in chemistry

The best exam tip is to know all your chemistry well! But however much (or little!) you do know, you wil do better if you have good exam technique and know what the examiner is looking for.

Be prepared

Make sure you know the style of the papers you will be taking. For example, will the paper be multiple-choice, structured (where you write answers in spaces on the paper) or free response (answers on lined paper)? How long will they be? Will they be broken down in some way – paper 1 on organic, paper 2 on physical, etc?

The best way to do this is to look at lots of past papers for the Board and syllabus that you will be sitting and check with your teacher that there have been no changes.

Read the question

This should go without saying, but do it carefully – especially where you have a choice of questions. It really is worth reading it twice. It is very common to misread a question the first time under the particular pressure of an exam. Also, you don't want to find out that the organic question that you leapt into was really about reaction rates when you've already spent 20 minutes on it.

The introduction to the question sometimes contains key words that could help with the answer. Be on the look-out for this.

Allocate your time

Most exam questions have a gradient of difficulty – that is, they start easily and get harder. So if you are short of time it makes sense to do the first (easy) parts of all the questions rather than struggle to finish just a few.

Know what the examiner means

Examiners have a language of their own and it is worth learning just what they mean by words such as 'state', 'describe' or 'explain'.

Here is a guide to what examiners mean by some commonly used terms.

Define means a formal statement such as 'A catalyst is a substance that alters the rate of a chemical reaction without itself being chemically changed in the process.'

What do you understand by? or **What is meant by?** means that you need to give a definition and some comment on the significance of the term in question. So you might add that a catalyst works by allowing a reaction pathway of lower activation energy.

State means that a short answer is required with no need for explanation. For example 'The oxidation number of oxygen in water is $-$II.'

List means give a number (usually specified) of points with no need for explanation. Do not give more points than are asked for.

Explain requires you to give some reasoning or refer to theory.

Describe means state in words (and possibly diagrams) the main points of a topic. If it refers to an experiment, you should say what you would see.

Discuss implies that you should give a critical account, pointing out pros and cons, for example.

Outline indicates a brief answer covering essential points only.

Predict or **deduce** means that you are expected to produce the answer by making logical connections between other pieces of information – probably ones mentioned (or worked out) earlier in the question.

Comment is open-ended, suggesting that you need to state or work out a number of points of interest and relevance to the context of the question. Exactly how long your comments are will depend on the marks, time and space allocated to that part of the question.

Suggest implies that there is no single 'right' answer. Either there may be several suitable answers or you may be expected to apply general knowledge or chemical reasoning to an unfamiliar situation.

Calculate or **determine** means that a numerical answer is needed. You should show working.

Estimate means give an answer of the right general size, possibly by making assumptions or rounding off quantities to simplify a calculation. For example you might be asked to estimate the fifth ionisation energy of an element having been given the first four.

Sketch, applied to a graph, means that the shape and position of the curve need to be correct but that actual values need not be. However, points such as passing through the origin (or not) are important.

In diagrams, 'sketch' means that a simple, freehand drawing is acceptable although the relative sizes of parts of an apparatus are important and the diagram should be clear.

This list has been adapted from one produced by the Cambridge Local Examinations Syndicate but the words will mean the same whatever the Board.

Give the examiner what (s)he wants

Look at the space for the answer on a structured paper. This will give you a clue as to how long your answer should be. If there are four lines and you think the answer is 'yes', then think again. Also look at the mark allocations for parts of questions. If there are three marks allocated, the examiner will be looking for three distinct points. On the other hand, if you need more space than is allocated for working out, use a sheet of paper and make sure it is included in your final script.

Keep the examiner happy

Examiners love signs, units, equations and state symbols. Always include units with your answer where appropriate. But do take care: not all numbers have units. Relative atomic mass, for example, is a pure number and it is wrong to give it units. Signs are important, especially in thermochemistry and electrochemistry. If a ΔH value is positive, put in a '+' sign. Wherever possible include balanced equations with state symbols.

Significant figures are significant

5 g is not the same as 5.00 g. In the first case we have weighed only to the nearest gram (to 1 significant figure) and in the second we know we have weighed to the nearest 0.01 g (to 3 significant figures).

The most important figure in any number and therefore the most significant is the first one from the left that isn't a zero. But, all the numbers that you write after the first one are called significant figures and they should match the accuracy in measurement.

- When doing a calculation, it is important that you don't just copy down the display of your calculator, as this may have far more significant figures than the data in the question justify. Our answer cannot be more accurate than the least accurate of the information we used.

For example, 81.0 g (3 significant figures) of iron has a volume of 10.16 cm^3 (4 significant figures). What is its density?

Density = mass/volume = 81/10.16 = 7.972 440 94 g cm^{-3} to 9 significant figures.

Since our least accurate measurement was to 3 significant figures, our answer should not be given beyond this and we should give the density to 3 significant figures, i.e. 7.97 g cm^{-3}.

If our answer had been 7.976 440 94, we should have rounded it up to 7.98 as the 4th significant figure is 5 or greater.

- The other point to be careful about is *when* to round up. This is best left to the very end of the calculation. Don't round up as you go along because it could make a difference to your final answer.

Calculator displays usually show numbers in standard form in a particular way. For example, the number 2.6×10^{-4} would appear as '2.6 − 04', a shorthand form which is not acceptable as an answer.

Use all the help you can get

In many A-level chemistry exams you will be given a copy of the Periodic Table and a Data Sheet or Data Book. One Board even allows you to take in a specified textbook for one paper. Make sure you know what you are given and how to use it. There is a Periodic Table in this book on p. 128, where you can find atomic number and atomic mass data.

How to use this book

Each chapter starts with a section 'What you should know now'. This tells you the basic chemistry that you must bring to the topic. This is followed by a selection of exam questions on that topic.

To save space, the questions in the book have the lines or spaces for the answers omitted but the marks allocated to each section of the question are given and are a useful indication of the length of answer required. Try the questions yourself and then check your answers and the commentaries when you have finished.

Some questions and answers have comments in with them:

Note This is to remind you of things that you should know but might miss.

Care This will alert you to aspects of the question where it is common to make mistakes.

Hint This will help you to make the most of the question by giving suggestions for the best way to tackle it.

Exam tip This will improve your chances of getting maximum marks.

1 PHYSICAL CHEMISTRY

What you should know now

❏ The terms:

rate of reaction

rate constant

rate expression

❏ [X] represents the concentration of species X in mol dm^{-3}.

❏ The number of moles of solute of concentration M mol dm^{-3} contained in V cm^3 of solution is given by

$$\text{Number of moles} = \frac{M \times V}{1000}$$

❏ The equilibrium law: reversible reactions, under appropriate conditions, produce an equlbirium mixture which contains reactants and products. For a reaction

$$aA + bB + cC\ldots \rightleftharpoons \ldots xX + yY + zZ$$

the expression $\quad K_c = \dfrac{[A]^a \times [B]^b \times [C]^c \ldots}{\ldots [X]^x \times [Y]^y \times [Z]^z}$

applies, where K_c is called the equilibrium constant (expressed in terms of concentration) at that particular temperature.

❏ Reactions with positive values of $E^\ominus$ are feasible.

❏ Acidities of solutions are expressed in terms of pH.

$$pH = -\log_{10}[H^+]$$

❏ Weak acids and bases are only partially dissociated into ions in solution. Strong acids and bases are fully dissociated.

❏ Le Chatelier's principle – that the position of an equilibrium reacts to reduce the effect of changes in conditions applied to it.

Exam questions

Q 1.1 A solution of hydrogen peroxide decomposes in the presence of a catalyst according to the equation:

$$2H_2O_2(aq) \rightarrow 2H_2O(l) + O_2(g)$$

In experiments to determine the rate of this reaction, hydrogen peroxide solutions of different concentrations were used with the same catalyst. The following results were obtained.

Experiment	$[H_2O_2(aq)]/$ $mol\,dm^{-3}$	Rate of reaction/ $mol\,dm^{-3}\,s^{-1}$
1	0.05	0.28×10^{-4}
2	0.15	0.85×10^{-4}
3	0.25	1.43×10^{-4}

(a) You are provided with $100\,cm^3$ of a $0.50\,mol\,dm^{-3}$ H_2O_2 solution, $10\,g$ of a solid catalyst and any necessary apparatus.
(i) How would you prepare $50\,cm^3$ of a $0.15\,mol\,dm^{-3}$ solution of H_2O_2 for use in experiment **2**?
(ii) Draw a labelled diagram of the apparatus you would use to carry out one of these experiments.
(iii) What measurements would you make in your experiments, and how would you use your results to obtain the rate of reaction? **8 marks**

(b) This decomposition is an example of a *disproportionation reaction*. Explain the meaning of this phrase by identifying the element undergoing disproportionation, and quote its oxidation numbers on both sides of the equation. **3 marks**

(c) (i) Plot the results of the experiments on the graph paper provided below.

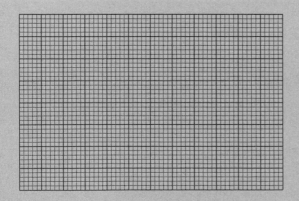

(ii) From the graph deduce the order of the reaction.
(iii) Hence write the rate equation for the reaction. **6 marks**

(d) This reaction is catalysed by solutions of transition metal ions. Explain why transition metal ions make good homogeneous catalysts, and explain how they work. **2 marks**

[L (Nuffield), '94]

Q 1.2 Chrome yellow, the pigment used for yellow road markings, is lead(II) chromate(VI), $PbCrO_4$.

(a) Write an equation, with state symbols, for the formation by precipitation of $PbCrO_4$. **1 mark**

(b) The solubility product of $PbCrO_4$ at $15\,°C$ is $1.69 \times 10^{-14}\,mol^2\,dm^{-6}$.

(i) Write an expression for the solubility product, K_{sp}, of $PbCrO_4$.
(ii) What is the solubility, in $mol\,dm^{-3}$, of $PbCrO_4$?
(iii) Concentrated aqueous lead(II) nitrate is added dropwise to $0.010\,mol\,dm^{-3}$ potassium chromate(VI). What is the concentration, in $mol\,dm^{-3}$, of lead(II) ions when the first trace of precipitate appears? **3 marks**

(c) Chrome yellow has been used for a long time as a yellow pigment in oil paintings.
(i) Use the standard redox potentials below to explain why the yellow colour changes when the painting is exposed to an atmosphere containing sulphur dioxide.

$$SO_4^{2-} + 4H^+ + 2e^- = 2H_2O + SO_2$$
$$E^\ominus = +0.17\,V$$
$$CrO_4^{2-} + 8H^+ + 3e^- = Cr^{3+} + 4H_2O$$
$$E^\ominus = +1.33\,V$$

(ii) Explain why this colour change takes a long time. **2 marks**

(iii) What colour change takes place? **1 mark**

[C, '95]

Q 1.3 Ethanoic acid is a weak acid, $pK_a = 4.76$.
(a) (i) Write an expression for the dissociation constant K_a of ethanoic acid.
(ii) Calculate the value of K_a, giving its units.
(iii) Calculate the pH of a solution of ethanoic acid of concentration $0.25\,mol\,dm^{-3}$. **5 marks**

(b) $20.0\,cm^3$ of sodium hydroxide solution of concentration $0.25\,mol\,dm^{-3}$ is slowly added to $10.0\,cm^3$ of ethanoic acid of the same concentration.
(i) Calculate the pH of the sodium hydroxide solution. (Ionic product of water $K_w = 1.0 \times 10^{-14}\,mol^2\,dm^{-6}$.)

(ii) Sketch on the axes below the expected change in pH as the sodium hydroxide is added to the ethanoic acid.

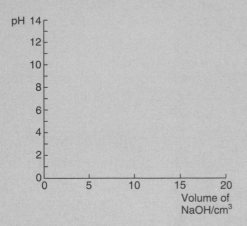

pH axis labelled from 0 to 14; horizontal axis labelled Volume of NaOH/cm³ from 0 to 20.

(iii) Show on your curve the pH *range* over which the mixture acts as a buffer.

(iv) Show, by means of an equation in each case, how this buffer solution responds to the addition of H^+ ions and OH^- ions.

(v) What volume of sodium hydroxide has to be added before the pH of the solution equals the pK_a of ethanoic acid? Briefly explain your reasoning.

(vi) Over what range of pH should the indicator in this titration change colour to obtain an accurate end-point? $\underline{\text{12 marks}}$

[L, '95]

1.4 (a) (i) Write down an expression for the ionic product of water, K_w, giving its units.
(ii) Define pH. $\underline{\text{3 marks}}$

(b) The value of K_w at various temperatures is given below:

Temperature/°C	0	50	100
K_w	0.114 $\times 10^{-14}$	5.48 $\times 10^{-14}$	51.3 $\times 10^{-14}$

(i) What is the pH of boiling water?
(ii) Is the dissociation of water exothermic or endothermic? Explain your answer. $\underline{\text{4 marks}}$

(c) Could the pH of water be altered by adding an insoluble catalyst at a fixed temperature? Explain your answer. $\underline{\text{3 marks}}$

(d) Explain in principle how a method based on the use of a hydrogen electrode might be used to determine the pH of water. No experimental details are required. $\underline{\text{3 marks}}$

[L, '93]

Answers

1.1 A solution of hydrogen peroxide decomposes in the presence of a catalyst according to the equation:

$$2H_2O_2(aq) \rightarrow 2H_2O(l) + O_2(g)$$

In experiments to determine the rate of this reaction, hydrogen peroxide solutions of different concentrations were used with the same catalyst. The following results were obtained.

Experiment	$[H_2O_2(aq)]/$ $mol\,dm^{-3}$	Rate of reaction/ $mol\,dm^{-3}\,s^{-1}$
1	0.05	0.28×10^{-4}
2	0.15	0.85×10^{-4}
3	0.25	1.43×10^{-4}

(a) You are provided with $100\,cm^3$ of a $0.50\,mol\,dm^{-3}$ H_2O_2 solution, $10\,g$ of a solid catalyst and any necessary apparatus.
(i) How would you prepare $50\,cm^3$ of a $0.15\,mol\,dm^{-3}$ solution of H_2O_2 for use in experiment **2**?

Using the expression that the number of moles of solute of concentration $M\,mol\,dm^{-3}$ contained in $V\,cm^3$ of solution is given by:

$$\text{Number of moles} = \frac{M \times V}{1000}$$

$50\,cm^3$ of $0.15\,mol\,dm^{-3}$ solution contains

$$\frac{50 \times 0.15}{1000} = 7.5 \times 10^{-3}\,\text{mol}$$

So we need to dilute to $50\,cm^3$ a volume of $0.50\,mol\,dm^{-3}$ solution which contains this number of moles.

$$7.5 \times 10^{-3} = \frac{0.50 \times V}{1000}$$

$$V = 15\,cm^3$$

So we must pipette out $15\,cm^3$ of the $0.50\,mol\,dm^{-3}$ solution into a $50\,cm^3$ volumetric flask and top it up to the mark with distilled water.

(ii) Draw a labelled diagram of the apparatus you would use to carry out one of these experiments

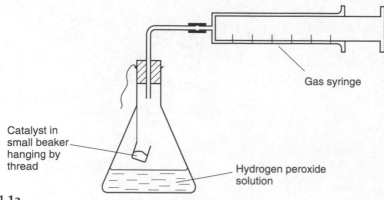

Gas syringe

Catalyst in small beaker hanging by thread

Hydrogen peroxide solution

Fig. 1.1a

The reaction can be started by shaking the flask to deposit the catalyst into the hydrogen peroxide solution.

The gas could equally be collected over water in an inverted measuring cylinder or burette.

(iii) What measurements would you make in your experiments, and how would you use your results to obtain the rate of reaction?

Take a series of readings of volume of oxygen in the gas syringe over time. Convert the volume of gas to moles. Plot a graph of number of moles of gas against time. The initial gradient (slope) of this graph is the rate of reaction at that particular concentration.

Note that we must use the *initial* gradient as that is the only time at which we know the concentration of the hydrogen peroxide solution exactly.

(b) This decomposition is an example of a *disproportionation reaction*. Explain the meaning of this phrase by identifying the element undergoing disproportionation, and quote its oxidation numbers on both sides of the equation.

Disproportionation describes a reaction in which the oxidation number of some atoms of a particular element goes up and that of others of the same element goes down.

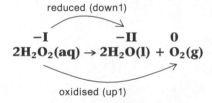

reduced (down1)

$$-I \qquad -II \qquad 0$$
$$2H_2O_2(aq) \rightarrow 2H_2O(l) + O_2(g)$$

oxidised (up1)

The oxidation number of oxygen in hydrogen peroxide (and other peroxides) is $-I$. This is unusual: oxygen's oxidation number in compounds is normally $-II$. Other exceptions are superoxides ($-\frac{1}{2}$) and compounds with fluorine where it is positive. In water it is the usual $-II$ and in O_2 (an uncombined element) zero. All the hydrogens have oxidation number $+I$.

So in this reaction, oxygen undergoes disproportionation – one atom being oxidised from $-I$ to 0 and the other being reduced from $-I$ to $-II$.

(c) (i) Plot the results of the experiments on the graph paper provided below.

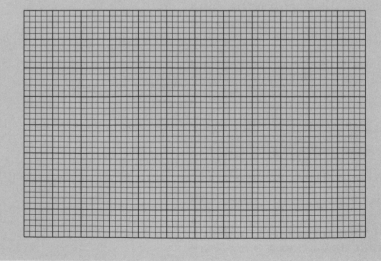

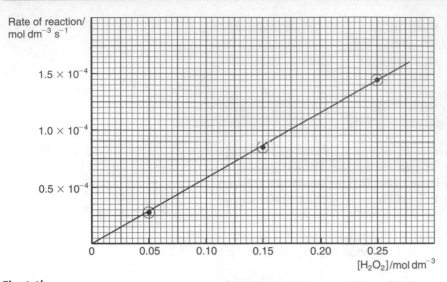

Fig. 1.1b

Remember to label the axes of the graph and include units. Choose a scale so that most of the graph paper is filled. The independent variable (concentration) goes on the horizontal axis.

(ii)   From the graph deduce the order of the reaction.

The graph of rate against [H$_2$O$_2$] is a straight line passing through the origin so rate is proportional to [H$_2$O$_2$] or [H$_2$O$_2$]^1. So the reaction is first order with respect to hydrogen peroxide concentration.

(iii)   Hence write the rate equation for the reaction.

Rate = k [H$_2$O$_2$]

where k is the rate constant.

The order of a reaction with respect to a particular reactant is the power to which the concentration of that reactant is raised in the rate expression. The overall order is the sum of the orders with respect to all the reactants.

The rate constant for a first-order reaction will have units of s^{-1}, the units of the rate divided by units of concentration. The rate constant for a second-order reaction would have units of dm^3 mol^{-1} s^{-1} – the units of rate divided by the units of concentration squared.

Remember that, unlike equilibrium constant expressions, rate expressions can only be worked out from experimental data – they are not related to the equation for the reaction.

(d)   This reaction is catalysed by solutions of transition metal ions. Explain why transitions metal ions make good homogeneous catalysts, and explain how they work.

Transition metal ions can usually display more than one oxidation number. They often work as catalysts by being oxidised by one of the reactants and then themselves oxidising some other species (or vice versa) so that they act as a temporary

reservoir for electrons. **This may create a new reaction pathway which has a lower activation energy than the uncatalysed reaction.**

Homogeneous catalysts are those which are in the same phase as the reactants – both in aqueous solution, for example. Heterogeneous catalysts are in a different phase from the reactants – solid catalysts and gaseous or aqueous reactants, for example.

WHY DOES IT MATTER?

Rocket planes

Rockets carry oxidising agents with which to burn their fuels. This means that they can fly at high altitudes where there is little or no atmospheric oxygen. The 1950s SR 53 rocket fighter prototypes burned kerosene fuel in oxygen produced by decomposition of concentrated hydrogen peroxide over a silver gauze – a heterogeneous catalyst. In 1958, one of the prototypes crashed and began to burn but the fire was rapidly extinguished by the water which was produced by the decomposition of the hydrogen peroxide.

1.2 Chrome yellow, the pigment used for yellow road markings, is lead(II) chromate(VI), $PbCrO_4$.
(a) Write an equation, with state symbols, for the formation by precipitation of $PbCrO_4$.

$Pb^{2+}(aq) + CrO_4^{2-}(aq) \rightarrow PbCrO_4(s)$

Since the question refers to precipitation, the ions must have started in aqueous solution. It does not matter where the ions came from, the lead might have been from lead nitrate and the chromate from potassium chromate, for example.

(b) The solubility product of $PbCrO_4$ at 15 °C is $1.69 \times 10^{-14} \, mol^2 \, dm^{-6}$.
(i) Write an expression for the solubility product, K_{sp}, of $PbCrO_4$.

$K_{sp} = [Pb^{2+}(aq)]_{eqm} \times [CrO_4^{2-}(aq)]_{eqm}$

This is an application of the general equilibrium law expression above applied to the equation in (a). This would give

$$K_c = \frac{[Pb^{2+}(aq)]_{eqm} \times [CrO_4^{2-}(aq)]_{eqm}}{[PbCrO_4(s)]_{eqm}}$$

However, it is impossible to change the concentration of a pure solid such as lead chromate (you cannot squeeze more of it into the same volume) and, by convention, its concentration is incorporated into the value of K_c. In the case of a sparingly soluble salt, the equilibrium constant, K_c, is called the solubility product, K_{sp}.

$$K_c \times [PbCrO_4(s)]_{eqm} = [Pb^{2+}(aq)]_{eqm} \times [CrO_4^{2-}(aq)]_{eqm}$$

$$K_{sp} = [Pb^{2+}(aq)]_{eqm} \times [CrO_4^{2-}(aq)]_{eqm}$$

CARE

Not all solubility products are as simple as this one. Remember that in equilibrium law expressions, the concentration of each species is raised to the power of its coefficient in the chemical equation. So for lead chloride, say,

$$Pb^{2+}(aq) + 2Cl^-(aq) \rightarrow PbCl_2(s)$$

$$K_{sp} = [Pb^{2+}(aq)]_{eqm} \times [Cl^-(aq)]_{eqm}^2$$

Solubility products can have different units. The lead chloride one has units of concentration cubed, i.e. $mol^3\,dm^{-9}$ and the lead chromate one units of concentration squared, i.e. $mol^2\,dm^{-6}$.

(ii) What is the solubility, in $mol\,dm^{-3}$, of $PbCrO_4$?

Solubility is the concentration of the salt in a saturated solution. The reverse of the equation in (a)

$$PbCrO_4(s) \rightarrow Pb^{2+}(aq) + CrO_4{}^{2-}(aq)$$

tells us that for every mole of $PbCrO_4(s)$ that dissolves we get one mole of dissolved $Pb^{2+}(aq)$ and one mole of dissolved $CrO_4{}^{2-}(aq)$. So the concentration of dissolved lead chromate = $[Pb^{2+}(aq)]_{eqm} = [CrO_4{}^{2-}(aq)]_{eqm}$.

$$K_{sp} = [Pb^{2+}(aq)]_{eqm} \times [CrO_4{}^{2-}(aq)]\ eqm$$

$$K_{sp} = [Pb^{2+}(aq)]_{eqm}{}^2$$

$$[Pb^{2+}(aq)]_{eqm} = \sqrt{K_{sp}}$$

$$[Pb^{2+}(aq)]_{eqm} = \sqrt{1.69 \times 10^{-14}}$$

$$[Pb^{2+}(aq)]_{eqm} = 1.3 \times 10^{-7}\,mol\,dm^{-3}$$

So the solubility of lead chromate is $1.3 \times 10^{-7}\,mol\,dm^{-3}$.

(iii) Concentrated aqueous lead(II) nitrate is added dropwise to $0.010\,mol\,dm^{-3}$ potassium chromate(VI).

What is the concentration, in $mol\,dm^{-3}$, of lead(II) ions when the first trace of precipitate appears?

The concentrations of $Pb^{2+}(aq)$ and $CrO_4{}^{2-}(aq)$ do not have to be equal – if we mixed solutions of lead nitrate and potassium chromate, for example. The solubility product tells us that if the product of the concentrations exceeds $1.69 \times 10^{-14}\,mol^2\,dm^{-6}$, then some solid lead chromate will precipitate.

$$K_{sp} = [Pb^{2+}(aq)]_{eqm} \times [CrO_4{}^{2-}(aq)]_{eqm}$$

$$1.69 \times 10^{-14} = [Pb^{2+}(aq)]_{eqm} \times 0.01$$

$$[Pb^{2+}(aq)]_{eqm} = 1.69 \times 10^{-14} \times 0.01$$

$$[Pb^{2+}(aq)]_{eqm} = 1.69 \times 10^{-16}\,mol\,dm^{-3}$$

(c) Chrome yellow has been used for a long time as a yellow pigment in oil paintings.
(i) Use the standard redox potentials below to explain why the yellow colour changes when the painting is exposed to an atmosphere containing sulphur dioxide.

$$SO_4{}^{2-} + 4H^+ + 2e^- = 2H_2O + SO_2 \qquad E^{\ominus} = +0.17\,V$$

$$CrO_4{}^{2-} + 8H^+ + 3e^- = Cr^{3+} + 4H_2O \qquad E^{\ominus} = +1.33\,V$$

This type of question can be approached in two ways. One is to use the 'anticlockwise rule'. Write the half-equations with electrons on the left and arrange them with the least positive value of $E^{\ominus}$ on top (as in the question).

The anticlockwise rule says that the top half-reaction will go to the left and the lower half-reaction to the right. So in this case SO_2 will turn into sulphate ions and yellow chromate ions will turn into green chromium(III) ions.

anticlockwise $\Bigg($
$$SO_4^{2-} + 4H^+ + 2e^- = 2H_2O + SO_2 \qquad E^{\ominus} = +0.17 \text{ V}$$
$$CrO_4^{2-} + 8H^+ + 3e^- = Cr^{3+} + 2H_2O \qquad E^{\ominus} = +1.33 \text{ V}$$

The other approach is to use the fact that reactions which have positive values of $E^{\ominus}_{reaction}$ are feasible. We work out $E^{\ominus}_{reaction}$ by adding the values of $E^{\ominus}$ for the two half-reactions. As both reactions are written with electrons on the left, one of the two reactions must be reversed so that we can add to two resulting equations and cancel out the electrons. In reversing the reaction, we must reverse the sign of $E^{\ominus}$.

This gives two possibilities:

A $\quad 2H_2O + SO_2 \rightarrow SO_4^{2-} + 4H^+ + 2e^- \qquad E^{\ominus} = -0.17 \text{ V}$

$\quad\quad CrO_4^{2-} + 8H^+ + 3e^- \rightarrow Cr^{3+} + 4H_2O \qquad E^{\ominus} = +1.33 \text{ V}$

$E^{\ominus}_{reaction} = +1.16 \text{ V}$ **and this reaction is feasible.** $E^{\ominus}$ **is greater than 0.6 V, which tells us that the reaction will go to completion.**

B $\quad SO_4^{2-} + 4H^+ + 2e^- \rightarrow 2H_2O + SO_2 \qquad E^{\ominus} = +0.17 \text{ V}$

$\quad\quad Cr^{3+} + 4H_2O \rightarrow CrO_4^{2-} + 8H^+ + 3e^- \qquad E^{\ominus} = -1.33 \text{ V}$

$E^{\ominus}_{reaction} = -1.16 \text{ V}$ **and this reaction is not feasible.**

Either method, using the anticlockwise rule or calculating $E^{\ominus}_{reaction}$, predicts that sulphur dioxide will reduce dichromate ions to chromium(III) ions.

NOTE
Even if $E^{\ominus}_{reaction}$ is positive, we can only predict that the reaction *might* occur. $E^{\ominus}_{reaction}$ tells us nothing about the rate of reaction, which could be very slow.

BY THE WAY

To get the correctly balanced form of the equation for the feasible reaction, the electrons must cancel. This means multiplying the upper equation by 3 and the lower one by 2 so that both involve 6 electrons.

$$6H_2O + 3SO_2 \rightarrow 3SO_4^{2-} + 12H^+ + 6e^- \qquad E^{\ominus} = -0.17 \text{ V}$$
$$2CrO_4^{2-} + 16H^+ + 6e^- \rightarrow 2Cr^{3+} + 8H_2O \qquad E^{\ominus} = +1.33 \text{ V}$$

Adding these gives

$$6H_2O + 3SO_2 + 2CrO_4^{2-} + 16H^+ + 6e^- \rightarrow 3SO_4^{2-} + 12H^+$$
$$+ 6e^- + 2Cr^{3+} + 8H_2O$$

Cancelling species which appear on both sides gives

$$3SO_2 + 2CrO_4^{2-} + 4H^+ \rightarrow 3SO_4^{2-} + 2Cr^{3+} + 2H_2O$$

NOTE
We do *not* multiply the values of $E^{\ominus}$ by 2 or 3; these are independent of the number of electrons involved.

(ii) Explain why this colour change takes a long time.

The concentration of sulphur dioxide in the air is small and so the reaction will be slow.

(iii) What colour change takes place?

Yellow (of chromate ions) to green (of Cr^{3+}).

This is the same colour change as in a chemical breathalyser, i.e. the reduction of Cr(VI) to Cr(III).

1.3 Ethanoic acid is a weak acid, $pK_a = 4.76$.
(a) (i) Write an expression for the dissociation constant K_a of ethanoic acid.

Ethanoic acid (CH_3CO_2H) dissociates partially in water as follows

$$CH_3CO_2H(aq) \rightleftharpoons H^+(aq) + CH_3CO_2^-(aq)$$

$$K_a = \frac{[H^+(aq)]_{eqm} \times [CH_3CO_2^-(aq)]_{eqm}}{[CH_3CO_2H(aq)]_{eqm}}$$

This is an example of the general equilibrium law expression.

(ii) Calculate the value of K_a, giving its units.

$$pK_a = -\log_{10} K_a = 4.76$$

$$\log_{10} K_a = -4.76$$

$$K_a = 1.73 \times 10^{-5} \, mol \, dm^{-3}$$

The best way to remember the expression for pK_a is by comparison with that for pH, i.e. 'p' means take the $\log_{10}$ of something and make it negative. To find K_a from $\log_{10} K_a$, use the INV $\log_{10}$ buttons on your calculator: punch in -4.76, then INV $\log_{10}$.

K_a has units of $mol \, dm^{-3}$ as it is a concentration squared divided by a concentration. pK_a (and pH) have no units as they are logs – the power to which ten must be raised to give K_a (or $[H^+]$).

(iii) Calculate the pH of a solution of ethanoic acid of concentration $0.25 \, mol \, dm^{-3}$.

Ethanoic acid dissociates partially as shown in the equation. Let us call $[H^+(aq)]_{eqm}$ x.

$$CH_3CO_2H(aq) \rightleftharpoons H^+(aq) + CH_3CO_2^-(aq)$$

start $0.25 \, mol \, dm^{-3}$ 0   0

at equilibrium $0.25 - x$ x x

$$K_a = \frac{x \times x}{(0.25 - x)}$$

$$1.73 \times 10^{-5} = \frac{x^2}{(0.25 - x)}$$

Since we are dealing with a weak acid, x is small and so $0.25 - x \approx 0.25$. So our expression becomes

$$1.73 \times 10^{-5} = \frac{x^2}{0.25}$$

$$x^2 = 1.73 \times 10^{-5} \times 0.25$$

$$x^2 = 4.325 \times 10^{-6}$$

$$x = 2.07966 \times 10^{-3}$$

$$[H^+(aq)]_{eqm} = 2.07966 \times 10^{-3} \, mol \, dm^{-3}$$

$$pH = 2.68$$

Whenever you do equilibrium calculations, it is vital to write out the chemical equation and the 'start' and 'at equilibrium' situations. In this case, since each ethanoic acid molecule which dissociates produces one H^+ and one $CH_3CO_2^-$, the equilibrium concentrations of these species must be equal. Since ethanoic acid is a weak acid, very few molecules actually dissociate, so the approximation that $[CH_3CO_2H]_{eqm} - [H^+]_{eqm} \approx [CH_3CO_2^-]_{start}$ is valid. This is true for any weak acid.

(b)   20.0 cm³ of sodium hydroxide solution of concentration 0.25 mol dm⁻³ is slowly added to 10.0 cm³ of ethanoic acid of the same concentration.
(i)   Calculate the pH of the sodium hydroxide solution. (Ionic product of water $K_w = 1.0 \times 10^{-14}$ mol² dm⁻⁶.)

$K_w = [H^+(aq)] \times [OH^-(aq)] = 1.0 \times 10^{-14}$ mol² dm⁻⁶ at 298 K.

In the sodium hydroxide solution, $[OH^-(aq)] = 0.25$ mol dm⁻³, as sodium hydroxide is a strong base.

$[H^+(aq)] \times 0.25 = 1.0 \times 10^{-14}$

$[H^+(aq)] = \dfrac{1.0 \times 10^{-14}}{0.25} = 4 \times 10^{-14}$ mol dm⁻³

pH = 13.4

In all aqueous solutions, the equilibrium

$$H_2O(l) \rightleftharpoons H^+(aq) + OH^-(aq)$$

is set up. The equilibrium law expression is

$$K_c = \frac{[H^+(aq)]_{eqm} \times [OH^-(aq)]_{eqm}}{[H_2O(l)]_{eqm}}$$

But as the concentration of pure, liquid water cannot change (we cannot squeeze any more into the same space), the term $[H_2O(l)]_{eqm}$ is incorporated into the equilibrium constant. So

$$K_c \times [H_2O(l)]_{eqm} = [H^+(aq)]_{eqm} \times [OH^-(aq)]_{eqm}$$

and we call the term $K_c \times [H_2O(l)]_{eqm}$ K_w, the ionic product of water.

(ii)   Sketch on the axes below the expected change in pH as the sodium hydroxide is added to the ethanoic acid.

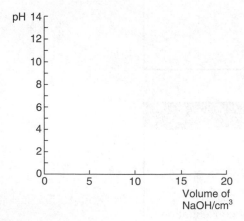

See Fig. 1.3a.

BY THE WAY

We can easily work out the concentration of liquid water from its density, 1 g cm⁻³. 1 dm³ (1000 cm³) of water has a mass of 1000 g. The relative molecular mass of water is 18, so 1 mol has a mass of 18 g. So 1 dm³ is 1000/18 = 55.5 mol and $[H_2O(l)]$ = 55.5 mol dm⁻³.

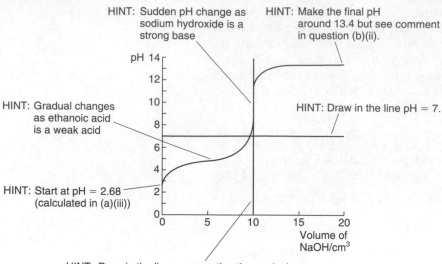

HINT: Sudden pH change as sodium hydroxide is a strong base

HINT: Make the final pH around 13.4 but see comment in question (b)(ii).

HINT: Gradual changes as ethanoic acid is a weak acid

HINT: Draw in the line pH = 7.

HINT: Start at pH = 2.68 (calculated in (a)(iii))

HINT: Draw in the line representing the equivalence point. It is at 10 cm³ because we have 10 cm³ of ethanoic acid which reacts 1:1 with sodium hydroxide of the same concentration.

Fig. 1.3a

The important points are marked on the graph.

It is important to realise that sodium hydroxide and ethanoic acid react 1:1, i.e.

$$NaOH(aq) + CH_3CO_2H(aq) \rightarrow CH_3CO_2Na(aq) + H_2O(l)$$

EXAM TIP

To be able to draw titration curves for any combination of strong or weak acid/strong or weak base, you need to learn the approximate shapes shown in Fig. 1.3b.

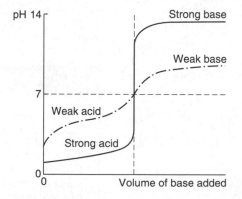

Strong base

Weak base

Weak acid

Strong acid

Volume of base added

Fig. 1.3b

BY THE WAY

The actual pH of the final solution which contains $10 \, cm^3$ of $0.25 \, mol \, dm^{-3}$ ethanoic acid and $20 \, cm^3$ of $0.25 \, mol \, dm^{-3}$ sodium hydroxide can be worked out.

Using the expression that the number of moles of solute of concentration $M \, mol \, dm^{-3}$ in $V \, cm^3$ of solution is

Number of moles $= \dfrac{M \times V}{1000}$

the solution contains

$$\dfrac{0.25 \times 10}{1000} = 2.5 \times 10^{-3}\, \text{mol}$$

ethanoic acid and

$$\dfrac{0.25 \times 20}{1000} = 5 \times 10^{-3}\, \text{mol}$$

sodium hydroxide.

As they react 1:1, the acid will neutralise $2.5 \times 10^{-3}\,$mol of the sodium hydroxide leaving $2.5 \times 10^{-3}\,$mol of sodium hydroxide dissolved in $30\,\text{cm}^3$ of solution. This gives a concentration of $2.5 \times 10^{-3} \times 1000/30\,\text{mol dm}^{-3} = 0.083\,\text{mol dm}^{-3}$ of sodium hydroxide. This has a pH of 12.9.

(iii) Show on your curve the pH *range* over which the mixture acts as a buffer.

The buffer region is the almost horizontal portion of the curve marked. Here, addition of either acid or base makes little change to the pH. See Fig. 1.3c.

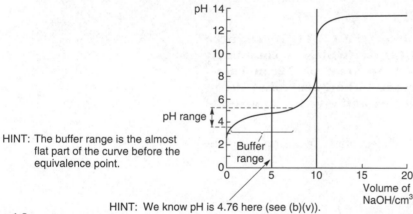

HINT: The buffer range is the almost flat part of the curve before the equivalence point.

HINT: We know pH is 4.76 here (see (b)(v)).

Fig. 1.3c

(iv) Show, by means of an equation in each case, how this buffer solution responds to the addition of H^+ ions and OH^- ions.

$$\mathbf{CH_3CO_2H(aq)} \;\rightleftharpoons\; \boxed{\begin{array}{c} \mathbf{H^+(aq)} \\ + \\ \mathbf{OH^-(aq)} \\ \downarrow \\ \mathbf{H_2O(l)} \\ \text{[2]} \end{array}} \;+\; \mathbf{CH_3CO_2^-(aq)} \quad \text{[1]}$$

Addition of H^+ ions moves equilibrium [1] to the left by Le Chatelier's principle and removes most of the added H^+ ions as undissociated CH_3CO_2H.

Added OH^- ions will react with H^+ ions via equation [2]. Equilibrium [1] will then move to the right by Le Chatelier's principle, i.e. more CH_3CO_2H will dissociate to try to restore the concentration of H^+ ions.

> (v)   What volume of sodium hydroxide has to be added before the pH of the solution equals the pK_a of ethanoic acid? Briefly explain your reasoning.

The Henderson–Hasselbach equation states that

$$pH = pK_a - \log_{10}\left(\frac{[CH_3CO_2H(aq)]}{[CH_3CO_2^-(aq)]}\right)$$

So for pH = pK_a,

$$\log_{10}\left(\frac{[CH_3CO_2H(aq)]}{[CH_3CO_2^-(aq)]}\right) = 0$$

Taking antilogs

$$\frac{[CH_3CO_2H(aq)]}{[CH_3CO_2^-(aq)]} = 1$$

So

$$[CH_3CO_2H(aq)] = [CH_3CO_2^-(aq)]$$

This can only be the case if half of the CH_3CO_2H has been neutralised and turned into CH_3CO_2Na (which is completely dissociated into Na^+ and $CH_3CO_2^-$). So $5\,cm^3$ of $0.25\,mol\,dm^{-3}$ sodium hydroxide solution are needed to half-neutralise $10\,cm^3$ of $0.25\,mol\,dm^{-3}$ ethanoic acid solution and give the solution of pH = pK_a ethanoic acid.

We can derive the Henderson–Hasselbach equation by rearranging the expression for K_a above.

$$K_a = \frac{[H^+(aq)]_{eqm} \times [CH_3CO_2^-(aq)]_{eqm}}{[CH_3CO_2H(aq)]_{eqm}}$$

Taking logs (and remembering that *adding* logs is the equivalent of multiplying),

$$\log_{10}K_a = \log_{10}[H^+(aq)]_{eqm} + \log_{10}\left(\frac{[CH_3CO_2^-(aq)]_{eqm}}{[CH_3CO_2H(aq)]_{eqm}}\right)$$

Making all the terms negative,

$$-\log_{10}K_a = -\log_{10}[H^+(aq)]_{eqm} - \log_{10}\left(\frac{[CH_3CO_2^-(aq)]_{eqm}}{[CH_3CO_2H(aq)]_{eqm}}\right)$$

$$pK_a = pH - \log_{10}\left(\frac{[CH_3CO_2^-(aq)]_{eqm}}{[CH_3CO_2H(aq)]_{eqm}}\right)$$

$$pH = pK_a + \log_{10}\left(\frac{[CH_3CO_2^-(aq)]_{eqm}}{[CH_3CO_2H(aq)]_{eqm}}\right)$$

Changing the sign of the log term is equivalent to inverting the fraction, so

$$pH = pK_a - \log_{10}\left(\frac{[CH_3CO_2H(aq)]_{eqm}}{[CH_3CO_2^-(aq)]_{eqm}}\right)$$

> (vi) Over what range of pH should the indicator in this titration change colour to obtain an accurate end-point?

Within the vertical portion of the pH/volume curve, i.e. roughly in the range pH 7 to pH 10.

A suitable indicator for a titration must change colour sharply (over the addition of no more than one drop) at the equivalence point of the titration, i.e. when the same number of moles of base have been added as the number of moles of acid that we started with – in this case after the addition of $10\,cm^3$ of sodium hydroxide. Most indicators change colour over a pH range of about 2 units. So we must look for a vertical portion of the titration curve of more than 2 units which coincides with the equivalence point.

WHY DOES IT MATTER?

Rasputin: was he a superman or was it chemistry?

Gregory Rasputin, the 'mad monk' who dominated the life of the last Tsarina of Russia and helped to bring down the Romanov dynasty was reputed to have supernatural powers. He was eventually killed by Prince Yussoupov (the nephew of Nicholas II, the last Tsar of Russia) but only after surviving an attempt to poison him with cakes and wine laced with 'enough potassium cyanide to kill several men instantly'. Was this evidence of mystic powers?

Probably not. The answer is more likely to be to do with weak acids. In the presence of water vapour from the air, carbon dioxide forms the weak acid carbonic acid. This has a pK_a of 6.4.

$$H_2O(l) + CO_2(g) \rightleftharpoons H^+(aq) + HCO_3^-(aq)$$

This acid is strong engough to react with potassium cyanide, turning it into hydrogen cyanide (a gas, which would escape into the air) and a harmless white solid potassium hydrogencarbonate which looks to the eye just like potassium cyanide.

$$KCN(s) + H^+(aq) + HCO_3^-(aq) \rightarrow KHCO_3(s) + HCN(g)$$

The most likely explanation for Rasputin's survival is that Yussoupov's 'cyanide' was old and had reacted with the moist air.

1.4 (a) (i) Write down an expression for the ionic product of water, K_w, giving its units.

$$K_w = [\text{H}^+(\text{aq})][\text{OH}^-(\text{aq})]$$

Its units are mol^2 dm^{-6}.

The equilibrium

$$\text{H}_2\text{O}(\text{l}) \rightleftharpoons \text{H}^+(\text{aq}) + \text{OH}^-(\text{aq})$$

exists in all aqueous solutions. The equilibrium constant for this is given by

$$K_c = \frac{[\text{H}^+(\text{aq})]_{\text{eqm}}[\text{OH}^-(\text{aq})]_{\text{eqm}}}{[\text{H}_2\text{O}(\text{l})]_{\text{eqm}}}$$

$[\text{H}_2\text{O}(\text{l})]$ cannot change so, by convention, its value ($55.55\,\text{mol dm}^{-3}$) is incorporated into the value of the equilibrium constant. This new equilibrium constant is called K_w and has the value of $10^{-14}\,\text{mol}^2\,\text{dm}^{-6}$ at $298\,\text{K}$. Its units are those of concentration squared. See **1.3** for more details.

(ii) Define pH.

pH $= -\log_{10}[\text{H}^+(\text{aq})]$

Temperature/°C	0	50	100
K_w	0.114×10^{-14}	5.48×10^{-14}	51.3×10^{-14}

(i) What is the pH of boiling water?

At 100 °C, $K_w = 51.3 \times 10^{-14}\,\text{mol}^2\,\text{dm}^{-6}$.

So

$$51.3 \times 10^{-14} = [\text{H}^+(\text{aq})][\text{OH}^-(\text{aq})]$$

Since each molecule of water dissociates to produce one H$^+$(aq) and one OH$^-$(aq), then [H$^+$(aq)] = [OH$^-$(aq)] and we can write

$$51.3 \times 10^{-14} = [\text{H}^+(\text{aq})]^2$$

So

$$[\text{H}^+(\text{aq})] = \sqrt{51.3 \times 10^{-14}}$$

$$[\text{H}^+(\text{aq})] = 7.16 \times 10^{-7}\,\text{mol dm}^{-3}$$

Since pH $= -\log_{10}[\text{H}^+(\text{aq})]$,

$$\textbf{pH} = \textbf{6.14}$$

(ii) Is the dissociation of water exothermic or endothermic? Explain your answer.

Endothermic.

The figures in the table show that the equilibrium

$$\text{H}_2\text{O}(\text{l}) \rightleftharpoons \text{H}^+(\text{aq}) + \text{OH}^-(\text{aq})$$

moves to the right at higher temperature (i.e. the equilibrium constant is larger). Le Chatelier's principle states that when an

BY THE WAY

This might surprise you! At 100 °C water is still neutral as it has the same number of H$^+$ ions as OH$^-$ ions but there are more of both of them than at room temperature.

equilibrium is disturbed, its position moves to minimise the disturbance. So if the temperature is raised, the equilibrium will move so as to minimise the temperature increase, i.e. to absorb heat. So dissociation absorbs heat (is endothermic).

(c)   Could the pH of water be altered by adding an insoluble catalyst at a fixed temperature? Explain your answer.

No. Catalysts do not affect the position of an equilibrium, merely the speed at which equilibrium is attained.

(d)   Explain in principle how a method based on the use of a hydrogen electrode might be used to determine the pH of water. No experimental details are required.

To measure [H$^+$(aq)] with a hydrogen electrode requires a hydrogen electrode to be set up connected via a high-resistance voltmeter and salt bridge to a reference electrode, for example a Cu/Cu^{2+} half-cell. The solution of unknown pH (water in this case) is used as the electrolyte in the hydrogen electrode. The measured e.m.f. of the cell is related to the hydrogen ion concentration via the Nernst equation

$$E = E^{\ominus} + \frac{RT}{zF} \ln [H^+(aq)]$$

where:

E = the measured e.m.f. of the cell

$E^{\ominus}$ = the e.m.f. of the cell under standard conditions (i.e. [H$^+$(aq)] = 1 mol dm^{-3}, temperature = 298 K and pressure = 100 kPa)

R is the gas constant, 8.3 J K^{-1} mol^{-1}

T is the temperature in kelvin

z is the number of electrons involved in the electrode reaction (in this case 1, as H$^+$(aq) + e$^-$ → $\frac{1}{2}$ H$_2$(g).

Substituting the values gives

$$E = E^{\ominus} + 0.026 \ln[H^+(aq)]$$

If the reference electrode is a standard Cu/Cu^{2+} half-cell, $E^{\ominus} = 0.34$ V and the expression becomes

$$E = 0.34 + 0.026 \ln[H^+(aq)]$$

So [H$^+$(aq)] and hence pH can be found.

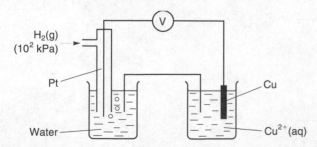

Fig. 1.4a

2 INORGANIC CHEMISTRY

What you should know now

❏ The colours of transition metal compounds are caused by light absorbed when electrons make transitions between d-orbitals.

❏ How to write down electron configurations of atoms and ions knowing the atomic number of the atom.

❏ How to calculate numbers of moles from masses of substances.

❏ Isomers are compounds of the same molecular formula but different arrangements of atoms in space.

❏ Some typical properties of transition elements.

Exam questions

2.1 (a) Explain why the hydrated copper(II) ion, e.g. in copper(II) sulphate pentahydrate, is coloured whereas copper(I) sulphate is white.   *3 marks*

(b) Illustrate the meaning of the term *disproportionation* by describing what happens when copper(I) sulphate reacts with water.   *3 marks*

(c) When 1.00 g of a copper(I) halide, CuX, reacted with aqueous 1,2-diaminoethane, $NH_2CH_2CH_2NH_2$, a blue solution and a precipitate were formed. On filtering this mixture, 0.22 g of copper metal was obtained.

By calculation, suggest the identity of X, and write an equation for the reaction.   *4 marks*

[C, '94]

2.2 (a) Describe briefly how, for an example of your own choice, the stoichiometry of a transition metal complex ion can be determined experimentally.   *4 marks*

(b) When aqueous ammonium dichromate(VI) is added gradually to melted ammonium thiocyanate, an ammonium salt known as Reinecke's salt, **A**, is formed. **A** has the formula $NH_4[Cr(SCN)_x(NH_3)_y]$, and analysis produced the following composition by mass:

Cr, 15.5%;
S, 38.1%;
N, 29.2%.

Calculate the values of x and y in the above formula and the oxidation number of chromium in the complex. Suggest a shape for the complex anion and explain what type of isomerism it exhibits.   *6 marks*

[C (International), '93]

Q 2.3 The following table has some information (items (i) to (vi)) omitted.

Species	Electron configuration	Number of unpaired electrons
Na^+	$1s^2 2s^2 2p^6$	(i)
Ti	(ii)	(iii)
(iv)	$1s^2 2s^2 2p^6 3s^2 3p^6 3d^5$	(v)
Cu	(vi)	one

(a) Give the information omited (items (i) to (vi)).   6 marks

(b) Which of the species in the completed table has the greatest paramagnetism? Explain your answer.   2 marks

(c) What is the maximum oxidation state titanium can show? Explain your answer, and give the formula of the nitrate of titanium in this oxidation state.   3 marks

(d) Give **two** large-scale uses of copper which depend on different properties of the metal. (State the uses and relevant properties.)   2 marks

(e) Give the name of **one** titanium compound of industrial importance, and indicate why it is used.   2 marks

[O, '94]

Answers

2.1 (a) Explain why the hydrated copper(II) ion, e.g. in copper(II) sulphate pentahydrate, is coloured whereas copper(I) sulphate is white.

The colour of transition metal ions is caused by electron transitions between one d-orbital and another. The electron configuration of a copper atom (29 electrons) is $1s^2$, $2s^2$, $2p^6$, $3s^2$, $3p^6$, $3d^{10}$, $4s^1$.

In copper(II), the 4s electron and one of the 3d electrons is lost giving an outer electron configuration of $3d^9$. The space in the d-orbitals means that electron transitions within the 3d orbitals are possible.

In forming a copper(I) ion, the 4s electron only is lost leaving a $3d^{10}$ arrangement. This has no empty d-orbitals, so there is no possibility of electron transitions within the d-orbitals and compounds containing this ion are normally colourless.

d-orbitals come in fives with the shapes and names shown in Fig. 2.1a. In an isolated atom, all five d-orbitals have exactly the same energy: we say that they are **degenerate**. However, when an atom is surrounded by ligands, this raises the energies of the d-orbitals. Orbitals whose lobes point directly towards the ligands have their energies raised more than those which do not. This raising of energy is caused by repulsion of the electrons in the d-orbitals by the electrons in the orbitals on the ligands. So in an octahedral complex ion such as the hydrated copper(II) ion, the energies of the $3d_{x^2-y^2}$ and the $3d_{z^2}$ orbitals are raised relative to the $3d_{xy}$, $3d_{yz}$ and $3d_{xz}$. See Fig. 2.1b. The energy gap between these orbitals corresponds to a quantum of electromagnetic radiation in the visible region of the spectrum. So the compound can absorb visible light and appears coloured.

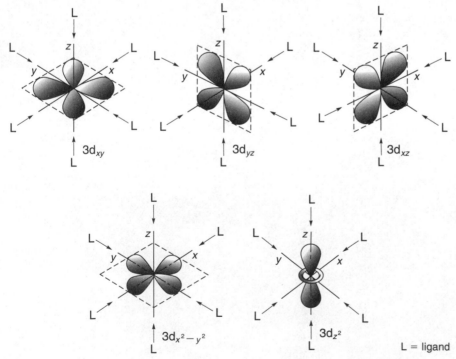

Fig. 2.1a

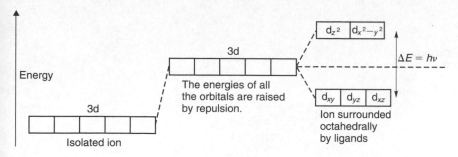

Fig. 2.1b

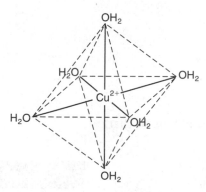

Fig. 2.1c   The Cu^{2+} ion is surrounded octahedrally by six water molecules acting as ligands.

> **NOTE**
>
> An octahedron has six *points* which is where the ligands are situated, the 'oct' part of the name comes from the fact that it has eight *faces* – see Fig. 2.1c.

(b)   Illustrate the meaning of the term *disproportionation* by describing what happens when copper(I) sulphate reacts with water.

Disproportionation describes a redox reaction in which an element in one oxidation state reacts to form products in two different oxidation states, one higher than the original and one lower. The element has been oxidised and reduced in the same reaction.

Copper(I) sulphate (which is white) dissolves in water to give a blue solution (copper(II) sulphate) and a pinkish coloured solid (copper – oxidation state 0). This is therefore a disproportionation reaction.

> **NOTE**
>
> The reverse of the process, where a higher and lower oxidation state of the same element react to form an intermediate oxidation state, is called (rather unimaginatively) reverse disproportionation.

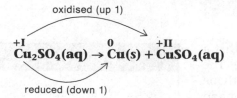

$$\overset{+I}{Cu_2SO_4}(aq) \rightarrow \overset{0}{Cu}(s) + \overset{+II}{CuSO_4}(aq)$$

oxidised (up 1)

reduced (down 1)

The $E^{\ominus}$/oxidation number diagram indicates that the equilibrium

$$Cu^{2+}(aq) + e^- \rightleftharpoons Cu^+(aq)$$

will move to the left and transfer electrons to the

$$Cu^+(aq) + e^- \rightleftharpoons Cu(s)$$

thus forcing the latter to the right. This is an example of the **anticlockwise rule**.

> **BY THE WAY**
>
> That this reaction will take place is shown by the redox potentials given in Fig. 2.1d.

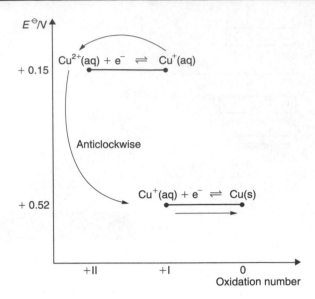

Fig. 2.1d

The overall reaction can be found by adding the two half-reactions so that the electrons cancel out.

$$Cu^+(aq) \rightarrow Cu^{2+}(aq) + e^- \qquad E^\ominus = -0.15\,V$$
$$\underline{Cu^+(aq) + e^- \rightarrow Cu(s)} \qquad \underline{E^\ominus = +0.52\,V}$$
$$\overline{2Cu^+(aq) + \cancel{e^-} \rightarrow Cu^{2+}(aq) + Cu(s) + \cancel{e^-}} \qquad \overline{E^\ominus = +0.37\,V}$$

Notice that as the direction of the top half-reaction is reversed compared with how it is written on the $E^\ominus$/oxidation number diagram, the sign of $E^\ominus$ is reversed. The value of $E^\ominus$ for the overall reaction (obtained by adding the $E^\ominus$s for the two half-reactions) is positive indicating that the reaction is feasible, though its value is less than $+0.6\,V$, indicating that it will not go to completion.

Even though a reaction is predicted to be feasible by use of $E^\ominus$ values, it may take place very slowly. $E^\ominus$ values tell us nothing about the rate.

> (c)   When 1.00 g of a copper(I) halide, CuX, reacted with aqueous 1,2-diaminoethane, $NH_2CH_2CH_2NH_2$, a blue solution and a precipitate were formed. On filtering this mixture, 0.22 g of copper metal was obtained.
>
> By calculation, suggest the identity of X, and write an equation for the reaction.

The formation of a blue solution and a copper precipitate suggests that a disproportionation reaction similar to the one above has taken place. The blue solution will be a complex of Cu^{2+} ions with 1,2-diaminoethane rather than $Cu^{2+}(aq)$. 1,2-diaminoethane is a better ligand than water and will tend to make the reaction go to completion. So the following reaction must be taking place.

$$2Cu^+(aq) \rightarrow Cu^{2+}(aq) + Cu(s)$$
2 mol   1 mol   1 mol

So if we obtain 0.22 g of copper, the original 1.00 g of copper halide must have contained twice this, 0.44 g of copper, leaving 0.56 g of halogen.

NOTE

The sulphate ions take no part in this reaction: they are spectator ions.

EXAM TIP

Notice how the previous section of the question gives a clue to this part.

0.44 g of copper is

   $0.44/63.5 = 6.93 \times 10^{-3}$ mol of Cu

As the formula of the halogen is CuX, it contains as many moles of halogen as of copper. So 0.56 g of halogen is 6.93×10^{-3} mol. So 1 mol of halogen X has a mass of $0.56/6.93 \times 10^{-3}$ g = 80.8 g, i.e. its relative atomic mass is 80.8. The only possible halogen is therefore bromine.

The equation for the reaction is

   $2CuBr(aq) + 2NH_2CH_2CH_2NH_2(aq) \rightarrow$

   $$Cu(NH_2CH_2CH_2NH_2)_2{}^{2+}(aq) + Cu(s) + 2Br^-(aq)$$

The actual formula of the copper/1,2-diaminoethane complex which forms depends on the concentration of the ligand, and three molecules of ligand are involved at higher concentration. Either answer would be acceptable.

WHY DOES IT MATTER?

Coloured transition metal compounds in everyday life

Transition metals are responsible for many of the colours we see around us in everyday life; different oxidation states of the same metal often have different colours. Lead chromate(VI) is the yellow pigment found in 'no-parking' lines while chromium(III) oxide is used in some green paints. The same chemistry was used in old fashioned breathalysers (now replaced by fuel-cell based ones). Yellow/orange crystals containing potassium dichromate(VI) were reduced to green chromium(III) by alcohol in a suspect's breath. The more crystals which turned green, the more alcohol in the breath.

Iron(III) oxide is used as a red pigment in some paints and titanium dioxide is the pigment in almost all white paints. The electron arrangement of a titanium atom is $1s^2, 2s^2, 2p^6, 3s^2, 3p^6, 3d^2, 4s^2$ but loss of four electrons in Ti^{4+} means that its electron arrangement is $3d^0$ and therefore it is colourless.

Further examples include cobalt compounds which give the blue of stained glass windows and traces of chromium compounds which give rubies their colour.

2.2 (a) Describe briefly how, for an example of your own choice, the stoichiometry of a transition metal complex ion can be determined experimentally.

The term 'stoichiometry' simply means the proportions or ratios in which species react to form compounds, so the stoichiometry of the complex ion $Cu(NH_3)_4{}^{2+}$ is 1 : 4.

There are a number of possible methods which would be acceptable as answers to this question. One of the most used (and easiest to describe) is the so called Job's method. It can be used in the very common situation where the complex ion is more intensely coloured than the ligands which react to form it. A suitable example would be the blood-red coloured complex formed between Fe^{3+} ions and thiocyanate ions (SCN^-). Solutions of these are mixed in varying ratios and the mixture with the darkest colour must be the one in which most complex has formed. This mixture must have the ligand and the metal ion mixed in the correct stoichiometric ratio so that there is neither excess ligand nor excess metal ion. A colorimeter with an appropriate filter to select light absorbed by the complex can be used to measure the amount of complex formed. The least amount of transmitted light means the maximum amount of complex.

An account like the one below should gain full marks.

To find the formula of the blood-red coloured complex formed between Fe^{3+} ions and thiocyanate ions (SCN^-), make two solutions, one containing Fe^{3+} (such as iron(III) chloride) and one containing SCN^- (such as potassium thiocyanate) of the same concentration. Set up a colorimeter with an appropriate filter to measure the colour of the complex. Make a series of mixtures with different proportions of metal ion and ligand solutions as shown in the table.

Mixture no.	1	2	3	4	5	6	7	8	9	10	11
Volume of Fe^{3+} solution	0	1	2	3	4	5	6	7	8	9	10
Volume of SCN^- solution	10	9	8	7	6	5	4	3	2	1	0
Colorimeter reading											

Plot a graph of the colorimeter reading against the composition of the mixture and extrapolate to find the composition of the mixture which transmits the minimum amount of light. This mixture must have the ligand and the metal ion mixed in the correct stoichiometric ratio so that there is neither excess ligand nor excess metal ion. In this case the stoichiometry of the complex is 1:1. 1:2 and 1:3 complexes would give the graphs shown in Fig. 2.2a.

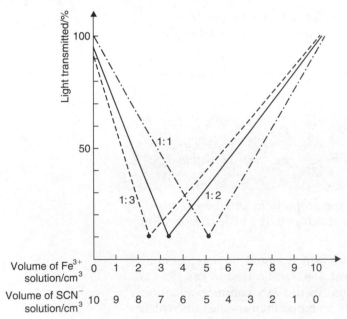

Fig. 2.2a

(b)   When aqueous ammonium dichromate(VI) is added gradually to melted ammonium thiocyanate, an ammonium salt known as Reinecke's salt, **A**, is formed. **A** has the formula $NH_4[Cr(SCN)_x(NH_3)_y]$, and analysis produced the following composition by mass:

Cr, 15.5%;

S, 38.1%;

N, 29.2%.

Calculate the values of x and y in the above formula and the oxidation number of chromium in the complex. Suggest a shape for the complex anion and explain what type of isomerism it exhibits.

First calculate the number of moles of each element in the complex by dividing the % composition by the mass of one mole (the relative atomic mass). Note that the percentages do not add up to 100% – the percentages of carbon and hydrogen have not been given.

Cr: 15.5/52 = 0.298 mol

S: 38.1/32 = 1.190 mol

N: 29.2/14 = 2.085 mol

These do not give a simple whole number ratio, so divide through by the smallest.

Cr: 0.298/0.298 = 1

S: 1.190/0.298 = 3.99

N: 2.085/0.298 = 6.99

This suggests a Cr:S:N ratio of 1:4:7.

Since sulphur is present only in SCN^- ions, there must be four SCN^- ions to each chromium atom. This accounts for four of the nitrogen atoms. One is present as an NH_4^+ ion so the other two must be present as NH_3 molecules.

This gives the formula as $NH_4[Cr(SCN)_4(NH_3)_2]$.

So $x = 4$ and $y = 2$.

As the charge on the ammonium ion is 1+, the charge on the complex ion $[Cr(SCN)_4(NH_3)_2]$ must be 1−. Ammonia molecules are neutral and the four thiocyanate ions add up to 4− so the oxidation state of the chromium atom must be +III.

The chromium atom in the complex ion is surrounded by six ligands (four SCN^- and two NH_3): the usual shape for this arrangement is octahedral – see Fig. 2.2b.

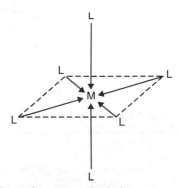

Fig. 2.2b An octahedral arrangement of ligands (L) around a metal ion (M).

NOTE

Oxidation states (or numbers) are usually given in Roman numerals.

Octahedral complexes of formula MX_4Y_2 show geometrical or *cis–trans* isomerism. The two Y ligands can be either adjacent or on opposite sides of the metal ion. See Fig. 2.2c.

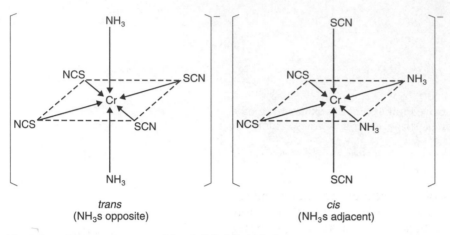

trans
(NH_3s opposite)

cis
(NH_3s adjacent)

Fig. 2.2c *Cis–trans* isomers of the $Cr(SCN)_4(NH_3)_2$ ions.

WHY DOES IT MATTER?

Cisplatin

Cisplatin is an anti-cancer drug whose formula is $Pt(NH_3)_2Cl_2$. It is believed to work by inhibiting the division of cancer cells by binding to their DNA. The geometry of the ligands around the platinum atom is square planar – flat and with all the bond angles 90°. Cisplatin is one of a pair of isomers. The *trans*-isomer, however, has no effect against cancer cells.

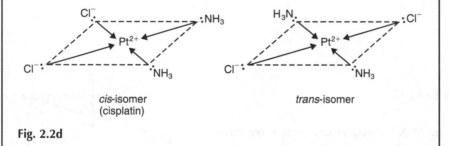

cis-isomer
(cisplatin)

trans-isomer

Fig. 2.2d

 2.3 The following table has some information (items (i) to (vi)) omitted.

Species	Electron configuration	Number of unpaired electrons
Na^+	$1s^2 2s^2 2p^6$	(i)
Ti	(ii)	(iii)
(iv)	$1s^2 2s^2 2p^6 3s^2 3p^6 3d^5$	(v)
Cu	(vi)	one

(a) Give the information omitted (items (i) to (vi)).

(i)   **None**

(ii) $1s^2 2s^2 2p^6 3s^2 3p^6 3d^2 4s^2$

(iii) **two (the two d-electrons)**

(iv) Mn^{2+} or Fe^{3+}

(v)   **five (the d-electrons)**

(vi) $1s^2 2s^2 2p^6 3s^2 3p^6 3d^9 4s^2$

An energy level diagram for all the atomic orbitals up to 4s is shown in Fig. 2.3a. Note that 4s is lower in energy than 3d. Electrons fill the levels from the lowest energy and work upwards (the **Aufbau principle**). When there is more than one orbital of the same energy (**degenerate orbitals**), they fill these separately at first and then pair up (**Hund's rule**).

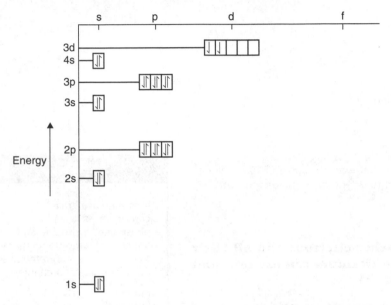

Fig. 2.3a Electron arrangement of a Ti atom. It has two unpaired electrons.

So in the case of a titanium atom with 22 electrons, 4s fills before 3d as it has lower energy and the electrons start to fill the five d-orbitals, separately at first so there are two unpaired d-electrons and the electron arrangement is as shown in Fig. 2.3a.

In the case of species (iv), there are five electrons in 3d but none in 4s. This means that the species must be an ion rather than a neutral atom. First-row transition metals lose their 4s electrons before their 3d electrons as they are further from the nucleus and therefore on the outside of the atom. Thus this species has lost both its 4s electrons. It is probably Mn^{2+}. A manganese atom would be $1s^2 2s^2 2p^6 3s^2 3p^6 3d^5 4s^2$, so loss of the two 4s electrons would give the electron arrangement in the table. Another possibility is Fe^{3+}. An iron atom is $1s^2 2s^2 2p^6 3s^2 3p^6 3d^6 4s^2$, so loss of three electrons would give the electron arrangement in the table – see Fig. 2.3b.

BY THE WAY

Species with the $3d^5$ arrangement have extra stability associated with their half-full d-shell. This is rather like the extra stability of species with full outer electron shells.

(b) Which of the species in the completed table has the greatest paramegnetism? Explain your answer.

Species (iv)

Paramagnetism is the property of being attracted into a magnetic field. It is associated with unpaired electrons so the species with the most unpaired electrons will show the greatest paramagnetism. Hence species (iv).

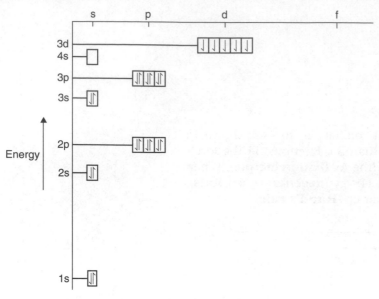

Fig. 2.3b The electron arrangement of Mn^{2+} and Fe^{3+}.

(c)   What is the maximum oxidation state titanium can show? Explain your answer, and give the formula of the nitrate of titanium in this oxidation state.

+IV

Transition metals can use both their s-electrons and all their unpaired d-electrons in bonding. So titanium can use four and show a maximum oxidation state of +IV.

Since the nitrate ion (NO_3^-) has a charge of -1, the formula of titanium(IV) nitrate will be $Ti(NO_3)_4$.

(d)   Give **two** large-scale uses of copper which depend on different properties of the metal. (State the uses and relevant properties.)

Copper is widely used for

- **electrical equipment and wiring as it has high electrical conductivity**
- **plumbing pipes as it is easy to work**
- **coinage as it is relatively unreactive**
- **heat exchangers as it conducts heat well.**

Any two of these answers would be acceptable. There is no point in giving more.

(e)   Give the name of **one** titanium compound of industrial importance, and indicate why it is used.

The most useful compound of titanium is titanium(IV) oxide, TiO_2, which is used as a white pigment for paint, plastics and paper.

WHY DOES IT MATTER?

Paint

Paint is a mixture of a resin (called a film-former), a pigment and a solvent. The pigment's role is to obscure the surface below the paint by scattering light and also to give the paint its colour. Titanium dioxide is used as a pigment in modern paints because of its good light scattering ability and its chemical unreactivity which prevents darkening associated with earlier lead-based pigments. This was caused by the reaction of the white basic lead carbonate with traces of hydrogen sulphide in the air to form black lead sulphide. Titanium dioxide is present in most coloured paints (except black) along with other transition metal oxides such as iron oxide (red) and chromium oxide (green) and also organic dyes.

The pigment is bonded to the surface by a resin film. In modern gloss paints the film is formed from alkyd (pronounced 'al kid') resins. These are polymers with branched chains. As the paint dries, the alkyd resins react with oxygen in the air and the chains cross link to form a 3-D network. The resins and the pigment are contained in a solvent such as white spirit which allows them to be applied by brush, roller or spray and then evaporates leaving behind the pigment and the resin film.

Paints can also contain additives to speed up drying, prevent settling in the can and to control the viscosity ('treacliness') of the paint. The latest paints are being developed to use water rather than white spirit as solvent because white spirit is flammable and may also help to form low level ozone in the atmosphere.

3 ORGANIC CHEMISTRY

What you should know now

❏ The reactions of halogenoalkanes – elimination and nucleophilic substitution.

❏ Isomers are compounds with the same molecular formula but different arrangements of the atoms in space.

❏ The reactions of aromatic rings.

❏ The reactions of alcohols, aldehydes and ketones, carboxylic acids, esters, alkyl and aryl amines.

❏ How the shapes of molecules are explained by electron pair repulsion theory.

Exam questions

3.1 (a) Give the reagents and conditions necessary to bring about the following changes for 2-bromobutane:

(i) $CH_3CH_2CHBrCH_3 \rightarrow CH_3CH_2CH = CH_2$
      A   **B**

(ii) $CH_3CH_2CHBrCH_3 \rightarrow CH_3CH_2CHOHCH_3$ <u>3 marks</u>
      A   **C**

(b) In part (a) (i) an alkene other than **B** is also formed.
(i) Identify this alkene.
(ii) State one type of isomerism shown by this compound which is not shown by **B**. Briefly explain how this occurs.   <u>3 marks</u>

(c) **A** can be converted into an acid **E** in a two-stage process via the compound **D**.

   $CH_3CH_2CHBrCH_3 \rightarrow D \rightarrow CH_3CH_2CH(CO_2H)CH_3$
       A   **E**

(i) Name compound **E**.
(ii) Give the structural formula of **D**.
(iii) What type of isomerism is shown by both **A** and **E**?
(iv) Give the reagents and conditions for the conversion of: **A** to **D**; **D** to **E**.   <u>7 marks</u>

(d) Give a mechanism for the conversion of **A** to **D**.   <u>3 marks</u>

[L, '95]

3.2 Each of the following concerns a pair of isomers. In each case write one possible structural formula for each of the species.

(a) **A** and **B** are alcohols with the molecular formula $C_4H_{10}O$. **A** turns warm acidified potassium dichromate(VI) solution green, but **B** does not.

(b) **C** and **D** have the molecular formula C_4H_8O and neither contains a carbon–carbon double bond. **C** reacts with ammoniacal silver nitrate but **D** does not.

(c) **E** and **F** have the molecular formula $C_3H_6O_2$. **E** liberates carbon dioxide from sodium hydrogencarbonate solution but **F** does not.

(d) **G** and **H** are amines with the molecular formula C_7H_9N and both contain a benzene ring. **G** forms a stable diazonium salt but **H** does not.

(e) **I** and **J** have the molecular formula C_4H_6. In a molecule of **I** all four carbon atoms lie in a straight line but in a molecule of **J** they do not.   <u>10 marks</u>

[N, '94]

Q 3.3 (a) In each case, name an organic group which gives

(i) an orange precipitate with 2,4-dinitrophenyl-hydrazine reagent

(ii) a red precipitate with Fehling's solution

(iii) copious fumes of hydrogen chloride with phosphorus pentachloride. _3 marks_

(b) The molecule **Q** below is the substance chiefly responsible for the smell of ripe raspberries.

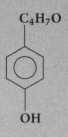

C_4H_7O

OH

It gives an orange precipitate with 2,4-dinitrophenylhydrazine reagent, no precipitate with Fehling's solution and no reaction with phosphorus pentachloride. It contains a chiral carbon atom.

(i) Draw the displayed formula of **Q**.

(ii) On your formula, draw a circle round the chiral carbon atom. _3 marks_

(c) (i) What would you expect to see when **Q** is treated with aqueous bromine?

(ii) Draw the displayed formula of the resulting organic product.

(iii) What type of reaction mechanism occurs when **Q** reacts with aqueous bromine? _3 marks_

[C, '95]

Answers

3.1 (a) Give the reagents and conditions necessary to bring about the following changes for 2-bromobutane:

(i) $CH_3CH_2CHBrCH_3 \rightarrow CH_3CH_2CH{=}CH_2$
       A   **B**

Heat with concentrated potassium (or sodium) hydroxide dissolved in ethanol.

(ii) $CH_3CH_2CHBrCH_3 \rightarrow CH_3CH_2CHOHCH_3$
        A   **C**

React with cold aqueous potassium (or sodium) hydroxide.

Note that the same reagent (potassium (or sodium) hydroxide) gives different products under different conditions. In (i) the OH^- ion acts as a base removing a proton followed by loss of a Br^- ion. This is an **elimination reaction**.

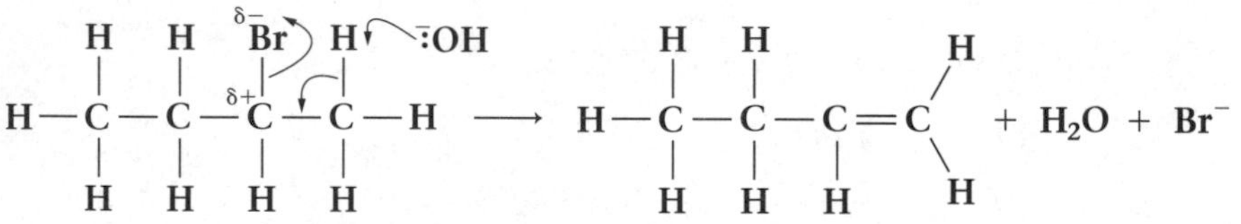

In (ii) the OH^- ion acts as a nucleophile, attacking the $C^{\delta+}$ to which the bromine atom is bonded and a **substitution** reaction occurs in which the bromine is lost as a Br^- ion. Nucleophilic substitution reactions can take place by two mechanisms – S_N1 and S_N2.

In S_N1, the C—Br bond breaks first by loss of a Br^- ion, leaving a carbocation (ion with a positively charged carbon atom) which is then attacked by OH^-. The slowest (rate-determining) step is the first and involves just one species.

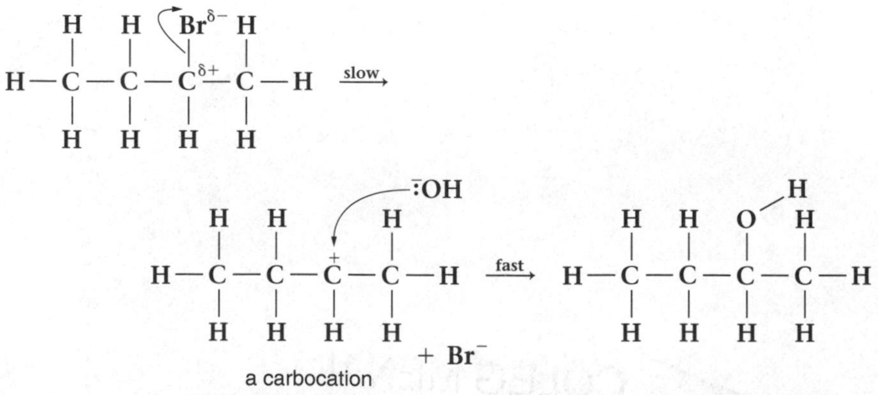

a carbocation

In S_N2, the OH^- ion attacks the $C^{\delta+}$, forming a bond by using its lone pair. This is the slowest step and involves two species. The resulting intermediate, in which the carbon is forming five bonds (three normal and two partial ones), rapidly breaks down by losing a Br^- ion – the leaving group.

EXAM TIP

It is often helpful to write out the full displayed formulae (showing every atom and every bond) of organic compounds even if it is not asked for. It helps to show the structure more clearly.

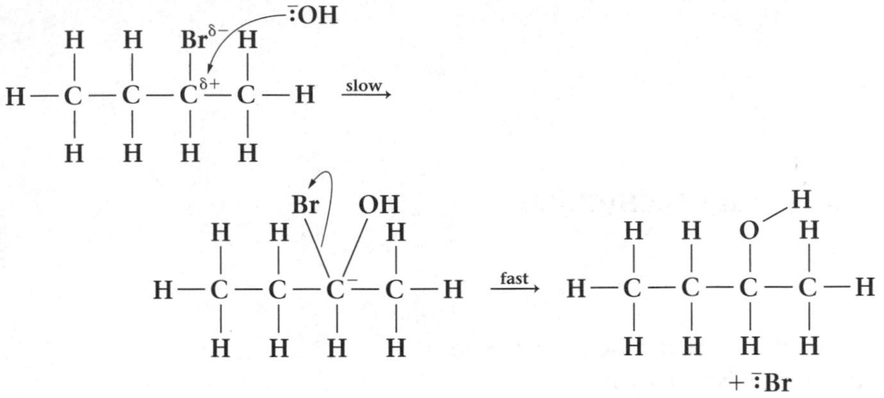

(b) In part (a) (i) an alkene other than **B** is also formed.
(i) Identify this alkene.

$CH_3CH{=}CHCH_3$. This is but-2-ene.

That shown in (a) (i) is but-1-ene.

The reaction takes place by the 2-bromobutane losing a proton followed by loss of the bromine as a Br^- ion. Two possible protons could be lost to start the process, resulting in two different products.

(ii) State one type of isomerism shown by this compound which is not shown by **B**. Briefly explain how this occurs.

***Cis–trans* isomerism (also called geometrical isomerism). This is where two substituents may be on the same (*cis*) or on opposite (*trans*) sides of a C=C.**

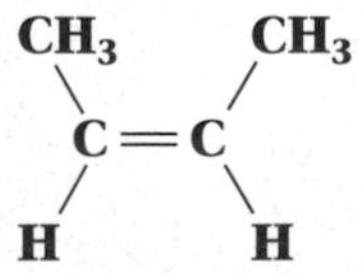

cis-but-2-ene

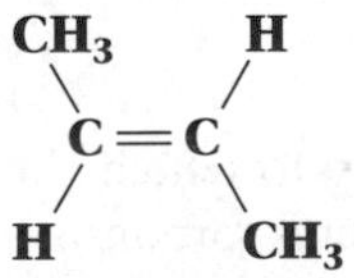

trans-but-2-ene

Remember that there is no rotation about double bonds.

(c) **A** can be converted into an acid **E** in a two-stage process via the compound **D**.

$CH_3CH_2CHBrCH_3$ (**A**) → **D** → $CH_3CH_2CH(CO_2H)CH_3$ (**E**)

(i) Name compound **E**.

E is 2-methylbutanoic acid.

Remember that the carbon of the $—CO_2H$ group is counted as part of the chain for naming purposes. The position of the substituent methyl group is counted from the functional group end of the carbon chain.

(ii) Give the structural formula of **D**.

D is

$$
H-\overset{\displaystyle H}{\underset{\displaystyle H}{C}}-\overset{\displaystyle H}{\underset{\displaystyle H}{C}}-\overset{\displaystyle \overset{N}{\parallel\!\parallel\!\parallel}}{\underset{\displaystyle H}{C}}-\overset{\displaystyle H}{\underset{\displaystyle H}{C}}-H
$$

or $CH_3CH_2CH(CN)CH_3$

D is a nitrile (2-methylbutanenitrile). The clue is that whenever we add one carbon atom in an organic reaction, a nitrile is usually involved.

(iii) What type of isomerism is shown by both **A** and **E**?

Optical isomerism.

$$
CH_3CH_2-\overset{\displaystyle Br}{\underset{\displaystyle H}{C^{*}}}-CH_3 \qquad\qquad CH_3CH_2-\overset{\displaystyle \overset{O}{\diagdown}\overset{}{C}\overset{O-H}{\diagup}}{\underset{\displaystyle H}{C^{*}}}-CH_3
$$

          A   **E**

Both **A** and **E** have a carbon atom (the chiral carbon, marked *) which is bonded to four different groups. Such molecules exist in two forms which are non-identical mirror images. These are called pairs of **enantiomers**.

Notice that you must consider the whole of the group to which the chiral carbon is bonded, not just the first atom. In **E** the chiral carbon is bonded to one H atom and three carbon atoms but these three are parts of different groups.

(iv) Give the reagents and conditions for the conversion of: **A** to **D**; **D** to **E**.

A to D: reflux with potassium cyanide in ethanol.
D to E: boil with an acid catalyst.

Nitriles are very useful in synthesis because they can add one carbon atom to a chain. They can also be converted into a variety of products. They can be converted (as here) into carboxylic acids which themselves can be made into alcohols, esters, amides, anhydrides and acid chlorides. They can also be reduced to give primary amines.

(d)   Give a mechanism for the conversion of **A** to **D**.

This is a nucleophilic substitution reaction.

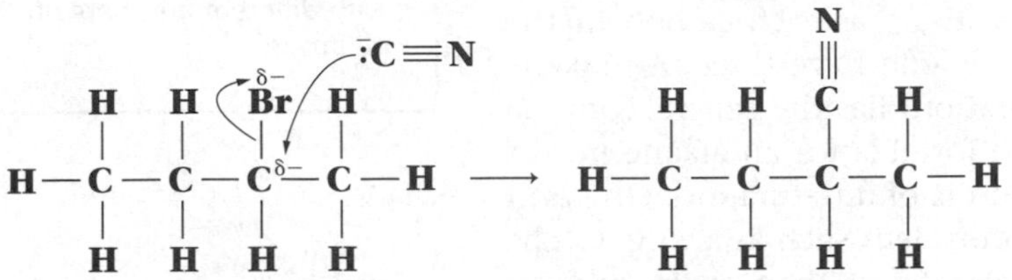

WHY DOES IT MATTER?

Organic synthesis

Organic synthesis is about designing methods to make a particular target compound from cheap, safe and readily available starting materials. It is very important in the pharmaceutical industry. One of the most common ways of discovering new drugs is to take a compound which is known to be pharmaceutically active – as a painkiller, antibiotic or anti-depressant, for example – and make compounds closely related to it – changing a hydrogen atom for a methyl group, replacing one halogen with another or adding a bulky group at a particular position, for instance. These new compounds are then tested for activity in 'test tube' experiments (*in vitro*) or on living organisms (*in vivo*) to try to establish so-called structure–activity relationships. These might be something like 'this group of atoms must remain intact for the compound to have any activity at all' or 'putting a more electronegative atom at this position makes it less toxic'. This will be followed by extensive testing and safety work. The attrition rate is high – typically only 1 in 10 000 compounds makes it from initial evaluation to sale as a drug.

BY THE WAY

During an S_N2 reaction, the nucleophile attacks from one side of the $C^{\delta+}$ and the leaving group leaves from the other. This means that the molecule 'turns itself inside out' rather like an umbrella in a high wind.

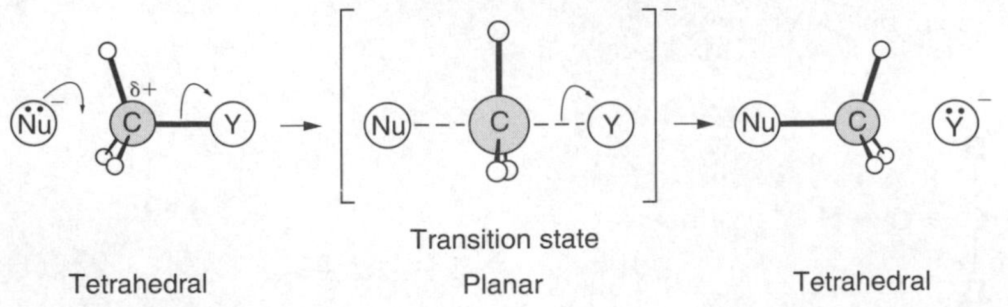

Tetrahedral   Planar   Tetrahedral

                        Transition state

3.2 Each of the following concerns a pair of isomers. In each case write one possible structural formula for each of the species.

The concept of **degree of unsaturation** is useful with this question and ones like it. An alkane has no double bonds – its degree of unsaturation is zero. The general formula of an alkane is C_nH_{2n+2} as each carbon in the chain has two hydrogens and the ones at the end have three. An alkene with one double bond (one unit of unsaturation) has the general formula C_nH_{2n}. Every two hydrogens fewer than the formula for an alkane means one unit of unsaturation. So C_4H_6 has two units of unsaturation. One unit of unsaturation is one double bond or its equivalent such as a ring. Cyclic hydrocarbons (i.e. rings) have two hydrogens fewer than chain ones as there are no end carbons. Two units of unsaturation could be two double bonds, one triple bond or a ring and a double bond. A double bond could be C=C, C=O, C=N and a triple bond C≡C or C≡N.

The idea can be used for compounds containing other elements as well as carbon and hydrogen. Halogens effectively replace hydrogens and so count as a hydrogen in calculating degree of unsaturation. So C_4H_7Br is the equivalent of C_4H_8 and has one unit of unsaturation. Oxygen would replace a CH_2 group (see Fig. 3.3a) and oxygens can therefore be ignored when calculating unsaturation, so C_2H_6O is equivalent to C_2H_6 and has no unsaturation.

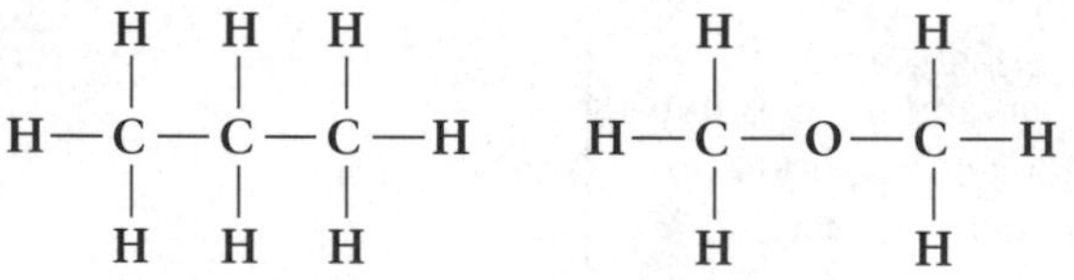

propane methoxymethane

An O replaces a CH_2 group.

Fig. 3.3a

To replace a —CH_2— group would require an —NH— group – see Fig. 3.3b. So to calculate the equivalent hydrocarbon formula, ignore the nitrogen and subtract one hydrogen. For example C_5H_9N is the equivalent of C_5H_8 and has two units of unsaturation.

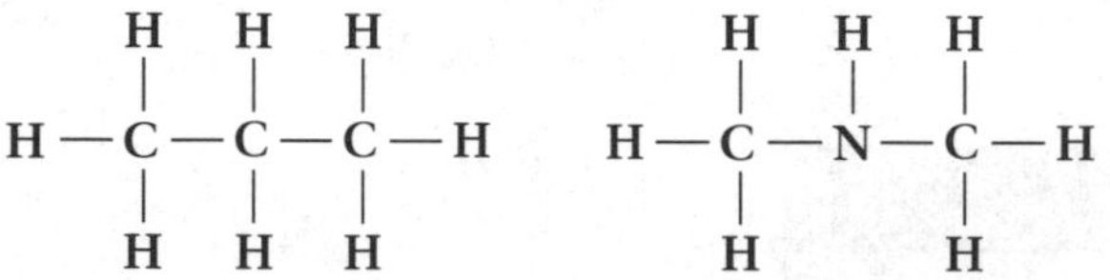

propane dimethylamine

An NH group replaces a CH_2 group.

Fig. 3.3b

(a) **A** and **B** are alcohols with the molecular formula $C_4H_{10}O$. **A** turns warm acidified potassium dichromate(VI) solution green, but **B** does not.

A is either

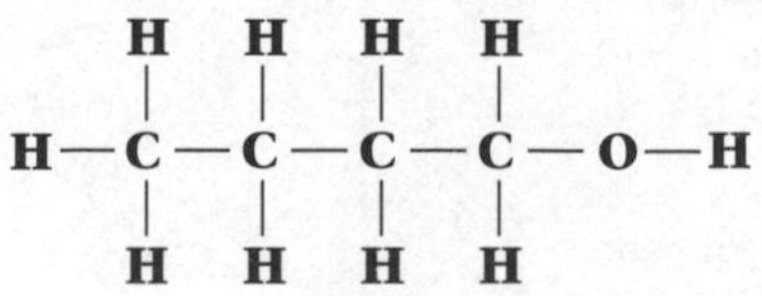

butan-1-ol (a primary alcohol)

CARE

Do not confuse isomers with isotopes which are atoms of the same element (i.e. with the same atomic number) but with different numbers of neutrons.

or

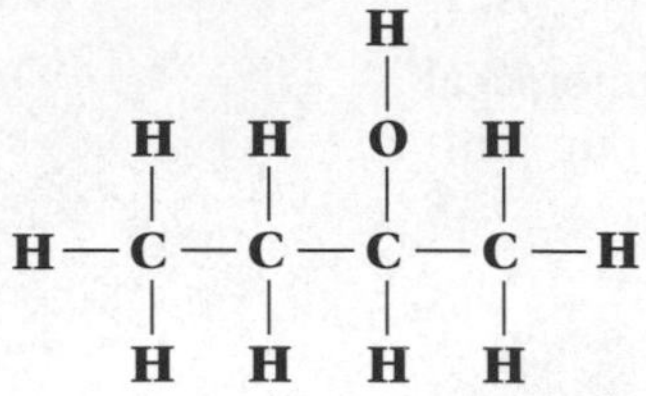

butan-2-ol (a secondary alcohol)

or

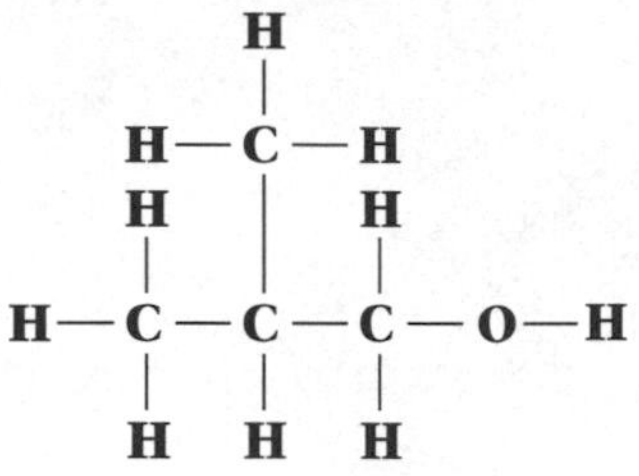

2-methylpropan-1-ol (a primary alcohol)

Any of these would gain full marks

B is

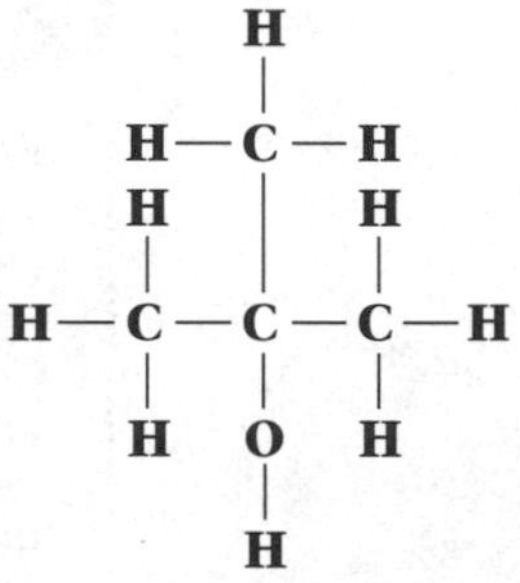

2-methylpropan-2-ol (a tertiary alcohol)

A is oxidised by warm acidified potassium dichromate solution so it must be a primary or a secondary alcohol. Primary alcohols are oxidised first to aldehydes and then to carboxylic acids. Secondary alcohols are oxidised to ketones and no further. Tertiary alcohols are not oxidised at all except under very vigorous conditions.

> **NOTE**
>
> Primary (1°) alcohols have the —OH group at the end of the hydrocarbon chain, secondary (2°) alcohols have the —OH in the middle of the chain and in tertiary (3°) alcohols the —OH group is at a branch in the chain.

(b)   **C** and **D** have the molecular formula C_4H_8O and neither contains a carbon–carbon double bond. **C** reacts with ammoniacal silver nitrate but **D** does not.

C is butanal

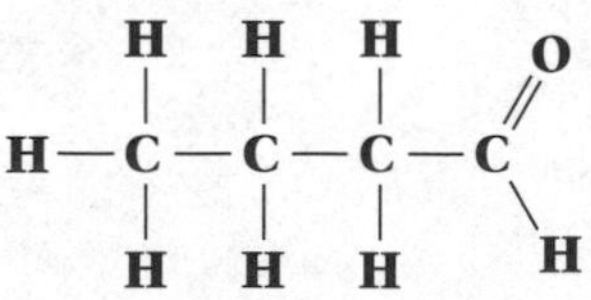

D could be a variety of isomers. The most obvious is butanone but tetrahydrofuran is another possibility.

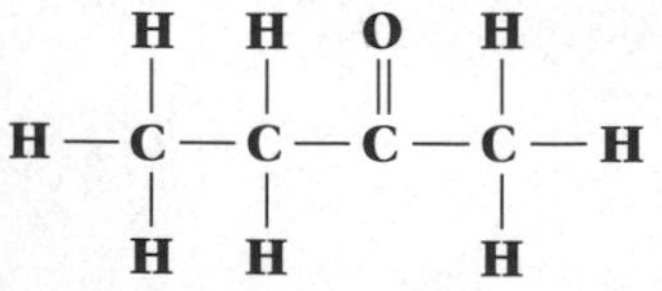

butanone

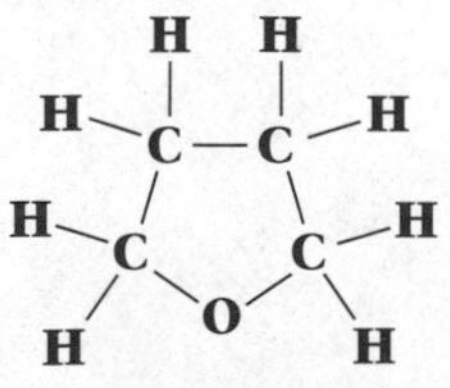

tetrahydrofuran

The formula C_4H_8 has one unit of unsaturation. As it is not a C=C, it could be a ring or a C=O. Formation of a silver mirror with ammoniacal silver nitrate is a specific test for an aldehyde – the functional group with C=O at the end of a chain.

> (c)   **E** and **F** have the molecular formula $C_3H_6O_2$. **E** liberates carbon dioxide from sodium hydrogencarbonate solution but **F** does not.

E is propanoic acid

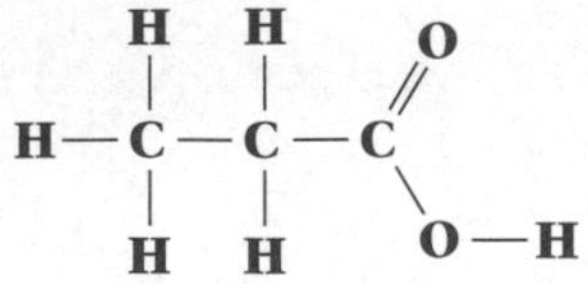

F is either methyl ethanoate

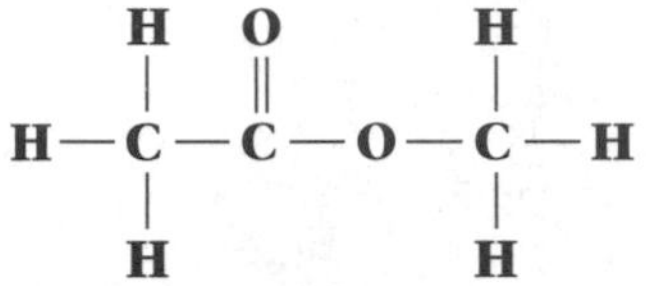

or ethyl methanoate

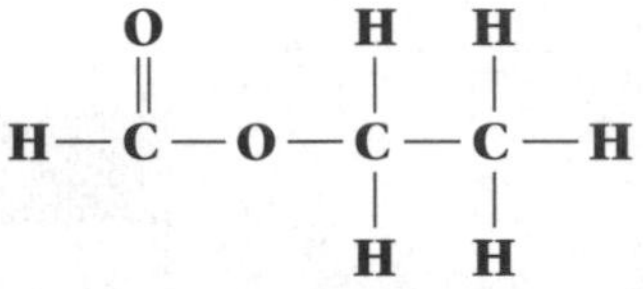

$C_3H_6O_2$ has one unit of unsaturation. **E**'s ability to liberate carbon dioxide from sodium hydrogencarbonate indicates that it is an acid and with three carbons propanoic acid is the obvious candidate. **F** is not acidic; a non-acidic isomer of a carboxylic acid suggests an ester. There are two possibilities.

EXAM TIP

It is easy to confuse the names and formulae of esters. Methyl ethanoate is derived from ethanoic acid, so draw the formula of ethanoic acid first and replace the hydrogen of the $-CO_2H$ group with a methyl group. Ethyl methanoate is derived from methanoic acid, so draw the formula of methanoic acid first and replace the hydrogen of the $-CO_2H$ group with an ethyl group.

> (d)   **G** and **H** are amines with the molecular formula C_7H_9N and both contain a benzene ring. **G** forms a stable diazonium salt but **H** does not.

G is 2-methylphenylamine

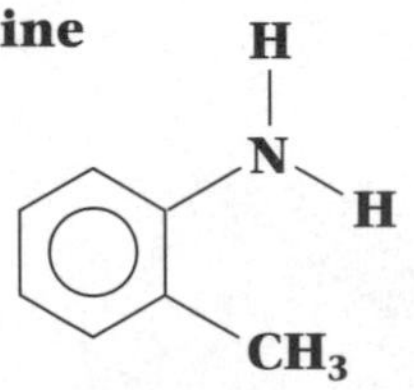

or 3-methylphenylamine

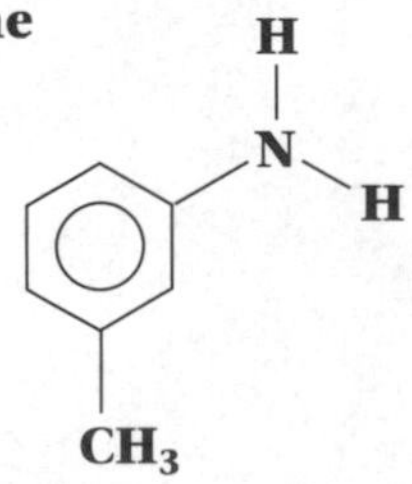

or 4-methylphenylamine

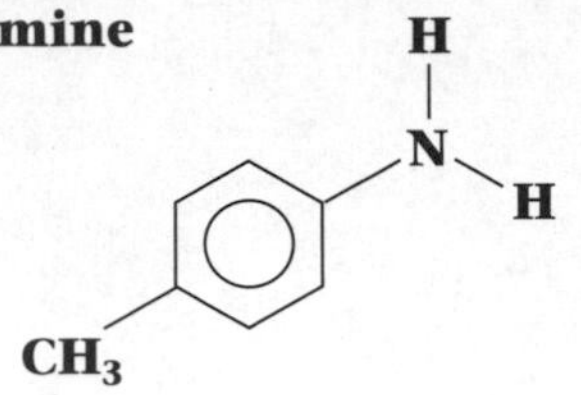

H is (phenylmethyl)amine

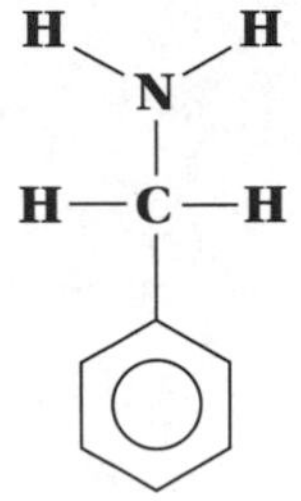

or N-methylphenylamine

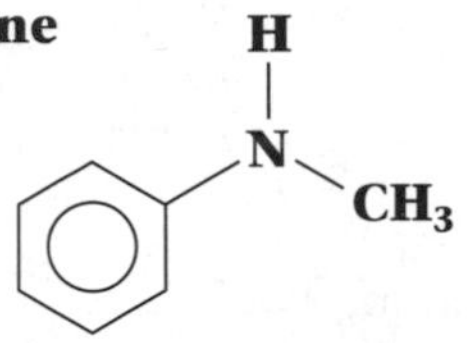

Primary aromatic amines form stable diazonium salts, so **G** must have an —NH_2 group attached directly to the benzene ring. This leaves a —CH_3 group unaccounted for and it could be in the 2-, 3- or 4-position on the ring.

H must be either an alkylamine like phenylmethylamine, where the amine group is not directly attached to the benzene ring, or a secondary aromatic amine like *N*-methylphenylamine. Notice how the names of these substances help to indicate their structure.

(e) **I** and **J** have the molecular formula C_4H_6. In a molecule of **I** all four carbon atoms lie in a straight line but in a molecule of **J** they do not.

I is

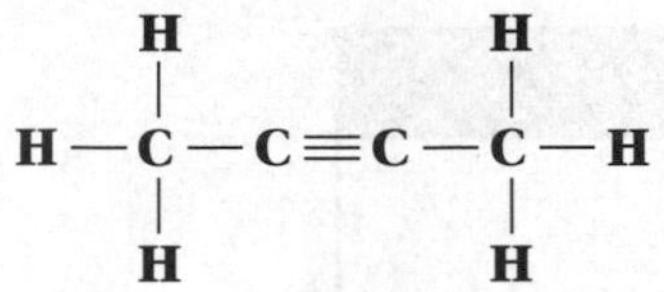

but-2-yne

J is

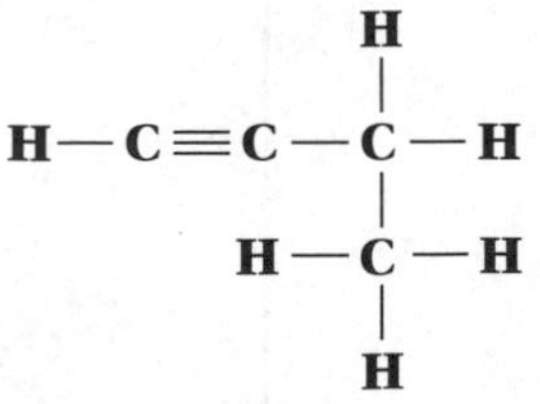

but-1-yne

or

$$
\underset{\text{buta-1,3-diene}}{
\begin{array}{c}
\text{H}\quad\quad\quad\text{H} \\
\text{H}-\text{C}\quad\text{C}=\text{C}-\text{H} \\
\text{C}=\text{C}\quad\quad\text{H} \\
\text{H}\quad\quad\text{H}
\end{array}}
$$

buta-1,3-diene

or

$$
\begin{array}{c}
\text{H}\quad\quad\quad\text{CH}_3 \\
\text{C}=\text{C}=\text{C} \\
\text{H}\quad\quad\quad\text{H}
\end{array}
$$

buta-1,2-diene

or

$$
\begin{array}{c}
\text{H}\quad\quad\text{H} \\
\text{C}=\text{C} \\
\text{H}-\text{C}-\text{C}-\text{H} \\
\text{H}\quad\text{H}
\end{array}
$$

cyclobutene

C_4H_6 represents two units of unsaturation – either a triple bond, two double bonds or a double bond and a ring. A look at the dot–cross diagram for but-2-yne shows two groups of electrons around each of the middle carbon atoms and hence a linear arrangement. Fig. 3.3c shows the orbitals involved – a σ orbital formed by overlap of two sp-orbitals and two π orbitals formed by overlap of p-orbitals.

π orbital

σ orbital

π orbital

$$
\begin{array}{c}
\text{H}\quad\quad\quad\quad\quad\text{H} \\
\text{H}-\text{C}\overset{\times}{\bullet}\text{C}\overset{\times\,\times}{\underset{\times\,\times}{}}\text{C}\overset{\times}{\bullet}\text{C}-\text{H} \\
\text{H}\quad\quad\quad\quad\quad\text{H}
\end{array}
$$

Fig. 3.3c The bonding in but-2-yne.

Why does it matter?

Dyes

Diazonium ions are made by the reaction of primary aromatic amines with nitrous acid (nitric(III) acid) which is generated by the reaction of sodium nitrite (sodium nitrate(III)) with hydrochloric acid. Phenylamine forms benzenediazonium chloride.

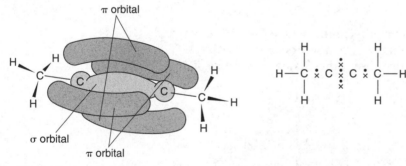

benzene diazonium chloride

Aryldiazonium ions are stable at low temperatures and can be made to react with phenols or aromatic amines in a reaction called coupling. This reaction joins the two aromatic systems through a —N=N— group so that the aromatic π systems are linked via the π-orbital of the —N=N— system – see Fig. 3.3d.

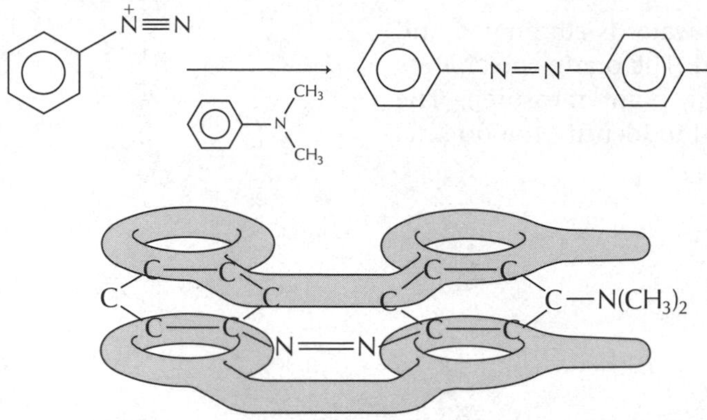

Fig. 3.3d The conjugated system formed after the coupling reaction.

This extended delocalised π system is described as conjugated. Conjugated systems make the molecules absorb light, so they are coloured. Many commercial dyes are based on such compounds. Two examples are shown below – Figs 3.3e and 3.3f.

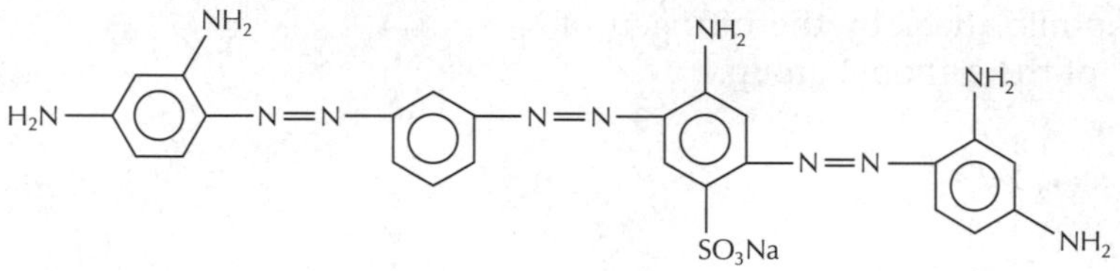

Fig. 3.3e C.I. Direct Brown 138.

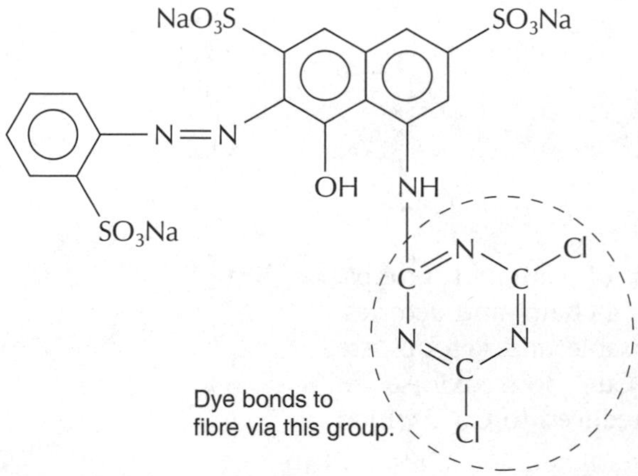

Fig. 3.3f Procion Brilliant Red 2BS, a reactive dye.

Direct dyes such as C.I. Direct Brown 138 bond to fabrics by weak van der Waals bonding and are therefore not very fast. Reactive dyes such as Procion Brilliant Red 2BS actually bond covalently to cotton or wool fabrics and will not wash out. They react via the group marked.

3.3 (a) In each case, name an organic group which gives
(i) an orange precipitate with 2,4-dinitrophenylhydrazine reagent.

The carbonyl group C=O, either an aldehyde or a ketone.

The reaction is an addition–elimination in which water is eliminated and an orange 2,4-dinitrophenylhydrazone is produced. The orange precipitate can be filtered off, recrystallised, and its melting point measured. The melting point can be looked up in tables and used to identify the original carbonyl compound.

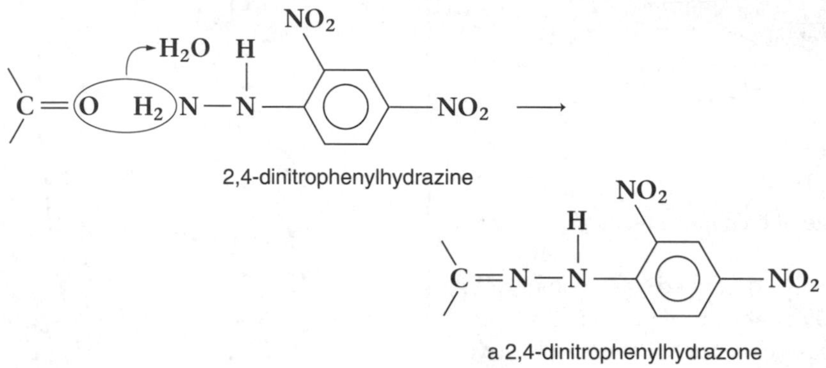

2,4-dinitrophenylhydrazine

a 2,4-dinitrophenylhydrazone

The initial step in the reaction is a nucleophilic attack by the nitrogen of the 2,4-dinitrophenylhydrazine on the $C^{\delta+}$ of the carbonyl group.

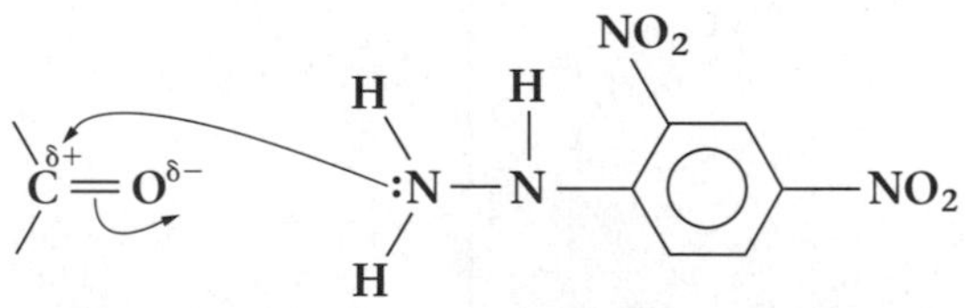

(ii) a red precipitate with Fehling's solution

An aldehyde, RCHO.

This test is used to distinguish the two types of carbonyl groups, aldehydes, RCHO, where the C=O is at the end of a chain, and ketones, R_2CO, where it is not. Aldehydes are easily oxidisable and ketones are not. In the Fehling's (or Benedict's) test, blue Cu^{2+} ions oxidise an aldehyde to a carboxylic acid and are themselves reduced to Cu^+, which is brick red.

(iii) copious fumes of hydrogen chloride with phosphorus pentachloride.

An —OH group either on its own (an alcohol) or as part of a carboxylic acid group —CO$_2$H but not one directly attached to a benzene ring (a phenol).

The reaction is

$$ROH + PCl_5 \rightarrow RCl + POCl_3 + HCl$$

BY THE WAY

Phenolic —OH groups do not react with phosphorus pentachloride because the basis of the reaction is a nucleophilic attack by a Cl⁻ ion on the $C^{\delta+}$ attached to the —OH group. In a phenol, this carbon is much less electron-deficient than in an alcohol because of the π electrons of the benzene ring which overlap with a p-orbital on the oxygen. So this carbon is less susceptible to nucleophilic attack and the C—O bond is stronger as it has some double bond character. See Fig. 3.2a.

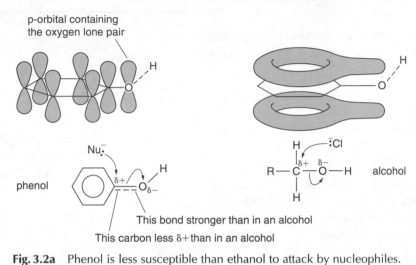

Fig. 3.2a Phenol is less susceptible than ethanol to attack by nucleophiles.

(b) The molecule **Q** below is the substance chiefly responsible for the smell of ripe raspberries.

C_4H_7O

OH

It gives an orange precipitate with 2,4-dinitrophenylhydrazine reagent, no precipitate with Fehling's solution and no reaction with phosphorus pentachloride. It contains a chiral carbon atom.
(i) Draw the displayed formula of **Q**.

Q is

(ii)   On your formula, draw a circle round the chiral carbon atom.

The chiral carbon is circled

The orange precipitate with 2,4-dinitrophenylhydrazine indicates that there is a carbonyl group present and the lack of reaction with Fehling's solution shows that this is a ketone, i.e. it is not at the end of a chain. Lack of reaction with phosphorus pentachloride shows that the C_4H_7O group does not contain an —OH group (the —OH on the benzene ring is a phenol and does not react).

The C_4H_7O group has one unit of unsaturation (see 3.3). This is accounted for by the C=O so there are no C=C's. This means that it must be based on

$$\begin{array}{ccccccc} & H & & H & & O & & H \\ & | & & | & & \| & & | \\ H - & C^4 & - & C^3 & - & C^2 & - & C^1 & - H \\ & | & & | & & & & | \\ & H & & H & & & & H \end{array}$$

The only ambiguity is the point of attachment of the benzene ring – at carbon 1, 3 or 4 in place of one hydrogen (carbon 2 is already forming four bonds). Attachment at carbons 1 and 4 would not give the molecule a chiral carbon atom while attachment at carbon 3 would.

Chiral carbon atoms

Some molecules exist as pairs of isomers (called enantiomers) which are non-identical mirror images of one another – see Fig. 3.2b. This occurs when there is a carbon atom in the molecule which is attached to four different groups. This carbon is called the chiral carbon or the stereo centre. The word chiral means 'handed' as in left- and right-handed. Enantiomers are virtually identical in their chemical properties, differing only in their effect on polarised light. One enantiomer will twist the plane of polarisation clockwise and the other anticlockwise.

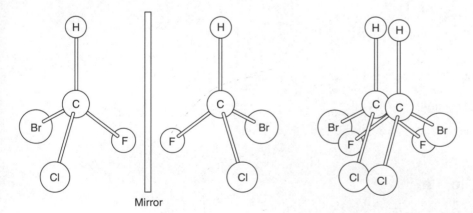

Fig. 3.2b   The molecule bromochlorofluoromethane exists as a pair of non-superimposable mirror images.

(c) (i)   What would you expect to see when **Q** is treated with aqueous bromine?

The bromine will slowly decolorise.

(ii)   Draw the displayed formula of the resulting organic product.

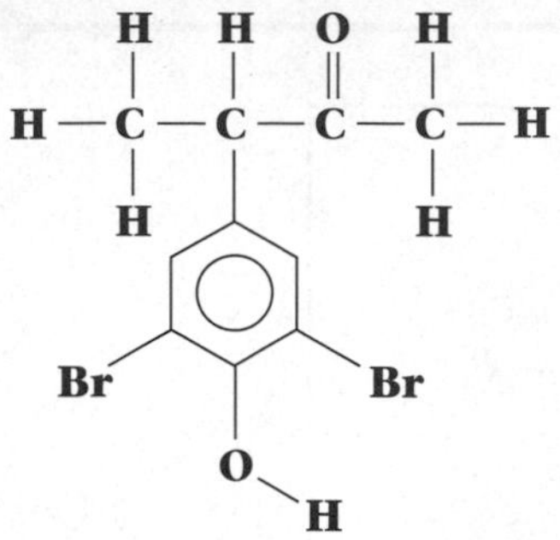

Bromine solution will react slowly with phenol via an electrophilic substitution reaction to give substitution at the 2-, 4- and 6-positions. In this case the 4-position is already blocked, so a 2,6-disubstituted product will be formed.

Aqueous bromine will not react with benzene itself – liquid bromine and an iron catalyst are required. The —OH group of phenol releases electrons into the benzene ring making it more prone to attack by electrophiles. So-called activating substituents normally activate the 2-, 4- and 6-positions.

(iii)   What type of reaction mechanism occurs when **Q** reacts with aqueous bromine?

An electrophilic substitution reaction.

Aromatic compounds normally react with electrophiles which are attracted by the high electron density of the π system of the ring. They normally undergo substitution rather than addition as addition would destroy the π system and cause the loss of the stability associated with aromatic compounds.

WHY DOES IT MATTER?

The optical isomers of thalidomide

In the late 1950s, the drug thalidomide was developed and used to treat the symptoms of morning sickness in pregnant women. By 1961 an increased number of children born with birth defects (often with no arms) was noticed and traced to the drug which was, of course, withdrawn.

Thalidomide was in fact not one substance but two – a pair of enantiomers (see Fig. 3.2c) – but the reactions used to make the drug produced a 50:50 mixture, called a **racemic mixture** or **racemate**. This was difficult to separate, so the drug was sold as a mixture. In fact only one of the isomers was active as a drug and the other caused the birth defects. The active form of thalidomide is now being used to treat leprosy, but not in women who might become pregnant, as the drug is known to racemise in the body.

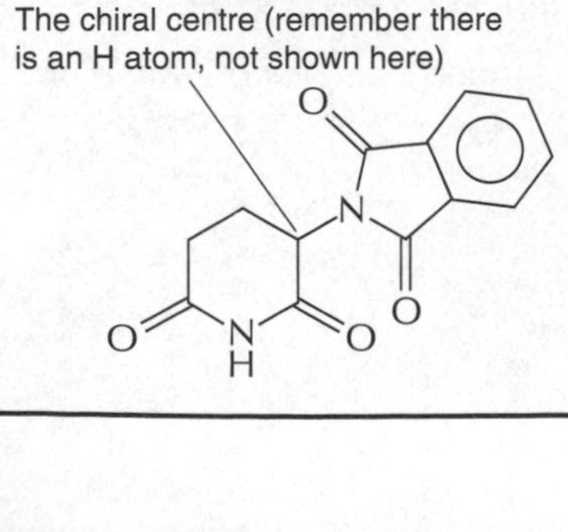

The chiral centre (remember there is an H atom, not shown here)

Fig. 3.2c

4 ENVIRONMENTAL CHEMISTRY

What you should know now

❑ How to draw dot–cross diagrams.

❑ How to calculate oxidation numbers of elements in compounds.

❑ How to do calculations involving moles of gases.

❑ How to write balanced equations for combustion reactions.

❑ Catalysts speed up chemical reactions without themselves being chemically changed. They work by reducing the activation energy for a reaction.

❑ Radicals and chain reactions.

❑ Rates of reaction/rate equations/order of reactions.

❑ Intermolecular forces.

❑ Tests for halide ions.

❑ Reactions of halogenoalkanes.

Exam questions

4.1 Hydrogen is a possible alternative to petrol as a fuel for cars in the future. One suggestion is to store the hydrogen in the car as magnesium hydride, MgH_2, and generate it as required by heating:

$$MgH_2 \rightarrow Mg + H_2$$

(a) (i) Give **one** advantage of using hydrogen in place of petrol as a fuel for motor cars.
(ii) Give **one** advantage of storing the fuel in the car in the form of magnesium hydride rather than hydrogen gas.   2 marks

(b) (i) Draw an electron dot–cross diagram to illustrate the bonding in magnesium hydride. Assume the bonding is purely ionic. Show the outer electrons only and indicate clearly the charge on each ion.
(ii) Explain why ionic coumpounds like magnesium hydride are crystalline solids.   4 marks

(c) One possible disadvantage of using magnesium hydride arises from its reaction with water:

$$MgH_2(s) + 2H_2O(l) \rightarrow Mg(OH)_2(s) + 2H_2(g)$$

(i) Explain why this could be a problem.
(ii) The reaction of magnesium hydride with water is a redox reaction. Using oxidation numbers, explain which element is being oxidised in the reaction and write a half-equation for the oxidation.   4 marks

(d) The fuel tank of one type of hydrogen-powered car holds 70 kg of MgH_2.

What *volume* of hydrogen gas, measured at room temperature and pressure, would be released if this amount of magnesium hydride reacted with water?

(Assume 1 mole of molecules of a gas at room temperature and pressure has a volume of 24 dm³.)

[Relative atomic masses: H = 1, Mg = 24]

[O&C (Salters') '93]

4.2 In petrol-driven cars, a mixture of hydrocarbons and air is ignited with a spark. The exhaust fumes from such cars contain various atmospheric pollutants such as NO, NO_2, CO and hydrocarbons. The pollutants can be removed from the exhaust fumes by use of a *catalytic converter* made from an aluminium alloy in the form of a fine mesh, coated with platinum or rhodium. The catalytic converter promotes reactions between molecules of the pollutant gases which result in less harmful exhaust fumes.

(a) (i) How do you account for the presence of carbon monoxide in exhaust fumes?
(ii) Write equations for the formation of the two oxides of nitrogen.
(iii) Suggest a reason why reactions involving nitrogen gas require a high temperature. _____ 5 marks

(b) (i) Explain how a catalyst affects the rate of a chemical reaction.
(ii) Why is the catalyst in the form of a fine mesh?
(iii) Why do you think that manufacturers recommend that cars fitted with catalytic converters should only use unleaded petrol?
(iv) Explain why a high temperature is necessary for the catalytic converter to operate. _____ 9 marks

(c) Suggest a method by which the concentrations of CO and hydrocarbons in exhaust fumes could both be reduced, other than by using catalysts. _____ 2 marks

[L, '94]

4.3 In 1930 the American Thomas Midgeley inhaled a lungful of the CFC with formula CCl_2F_2 and used it to blow out a candle. He did this to demonstrate a new refrigerant which was non-flammable and non-toxic.

(a) What does CFC stand for? _____ 1 mark

(b) State **two** other uses (as well as refrigerants) to which CFCs have been put. _____ 2 marks

(c) In the stratosphere, CFCs are broken down to give chlorine atoms. These chlorine atoms react with ozone molecules thus:

$$Cl^\bullet + O_3 \rightarrow ClO^\bullet + O_2 \qquad [1]$$

(i) Write *equation* [2] which shows how the $ClO^\bullet$ radicals can react further with oxygen atoms to give oxygen molecules and chlorine atoms. _____ 1 mark
(ii) Use *equations* [1] and [2] to give an overall equation. Then explain why a few chlorine atoms can cause the destruction of many ozone molecules. _____ 2 marks

(iii) The reaction represented by *equation* [1] is *first order* with respect to both chlorine atoms and ozone molecules. Write the *rate equation* for this reaction. _____ 1 mark

(iv) What happens to the rate of ozone depletion by chlorine atoms if the concentration of chlorine atoms is doubled? Explain your answer. _____ 2 marks

(v) Give the equation for a *termination* reaction in which chlorine atoms are removed from the stratosphere. _____ 1 mark

(vi) Why is the rate of this termination reaction slow in the stratosphere? _____ 1 mark

(d) If the production of CFCs were to be phased out now, it would be a long time before their damaging effects on the ozone layer began to be reduced. Suggest *two* reasons for this. _____ 2 marks

(e) Suggest a reason why Midgeley was unaware of the drawbacks associated with CFCs. _____ 1 mark

(f) Chemists are now looking for compounds to replace CFCs. Describe **two** properties which these compounds should have, as well as being non-toxic and non-flammable. _____ 2 marks

(g) A variety of halogenoalkanes similar to Midgeley's original compound have been synthesised to see whether they might be useful. For example, CF_2I_2 has been synthesised.

(i) Explain why this compound has a higher boiling point than CCl_2F_2. _____ 2 marks

(ii) Give a reason why CF_2I_2 reacts faster than CCl_2F_2 with sodium hydroxide solution. _____ 1 mark

(iii) When CF_2I_2 reacts with sodium hydroxide, iodide ions are formed in the alkaline solution. In order to test for the presence of these ions a suitable substance has first to be added to neutralise the alkali.

If you were doing this test, state:
the substance you would use to carry out the neutralisation; _____ 1 mark

the substance you would add to test for the iodide ions; _____ 1 mark

the positive result of the test. _____ 1 mark

[O&C (Salters') '94]

Answers

4.1 Hydrogen is a possible alternative to petrol as a fuel for cars in the future. One suggestion is to store the hydrogen in the car as magnesium hydride, MgH_2, and generate it as required by heating:

$$MgH_2 \rightarrow Mg + H_2$$

(a) (i) Give **one** advantage of using hydrogen in place of petrol as a fuel for motor cars.

When petrol burns, the products include

- **carbon dioxide (a greenhouse gas)**
- **sulphur dioxide (which contributes to acid rain) from sulphur in the fuel**
- **nitrogen oxides (which contribute to acid rain) from combination of atmospheric nitrogen and oxygen at the high temperature of the flame**
- **water.**

Burning hydrogen produces no carbon dioxide and no sulphur dioxide. It produces water and nitrogen oxides.

Note that it would not have been correct to make a statement such as 'burning hydrogen causes no pollution': nitrogen oxides will still be formed.

Another possible approach to this question would have been to say that hydrogen could be a renewable fuel whereas petrol (made from crude oil) is not. However, hydrogen is only renewable if it is made (by electrolysis of water) using a renewable source of energy such as electricity from sunlight or hydro power. At present, most hydrogen is made from fossil fuels.

(ii) Give **one** advantage of storing the fuel in the car in the form of magnesium hydride rather than hydrogen gas.

Hydrogen is a gas and is therefore bulky to store (even under pressure) and difficult to handle during refuelling. Solids such as magnesium hydride are easier to store and handle.

Your answer could also refer to the flammability of hydrogen.

(b) (i) Draw an electron dot–cross diagram to illustrate the bonding in magnesium hydride. Assume the bonding is purely ionic. Show the outer electrons only and indicate clearly the charge on each ion.

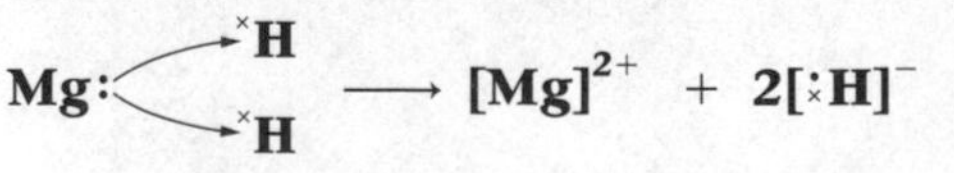

$$Mg : \overset{\displaystyle {}^{\times}H}{\underset{\displaystyle {}^{\times}H}{<}} \longrightarrow [Mg]^{2+} + 2[:H]^-$$

Fig. 4.1a

It would not be necessary to show the electron transfers to get full marks but you will probably find it helpful. Notice the hydride ion (H^-) is formed by *gain* of an electron by a hydrogen atom. Hydrogen more usually forms H^+ ions by *loss* of an electron.

(ii)   Explain why ionic compounds like magnesium hydride are crystalline solids.

The ions attract each other by electrostatic forces leading to a regular giant structure with alternating positive and negative ions. The electrostatic attractions extend throughout the structure.

(c)   One possible disadvantage of using magnesium hydride arises from its reaction with water:

$$MgH_2(s) + 2H_2O(l) \rightarrow Mg(OH)_2(s) + 2H_2(g)$$

(i)   Explain why this could be a problem.

The magnesium hydride would have to be kept dry during storage or flammable hydrogen gas would be formed causing a risk of explosion.

(ii)   The reaction of magnesium hydride with water is a redox reaction. Using oxidation numbers, explain which element is being oxidised in the reaction and write a half-equation for the oxidation.

$$\overset{+II\ -I}{MgH_2(s)} + \overset{+I\ -II}{2H_2O(l)} \rightarrow \overset{+II\ -II+I}{Mg(OH)_2(s)} + \overset{0}{2H_2(g)}$$

oxidised (up 1)

reduced (down 1)

MgH_2: Magnesium (and other Group 2 elements) *always* form oxidation number +II in their compounds. So the hydrogens in magnesium hydride must each be −I. Note that this is an exception; hydrogen is usually +I in its compounds.

H_2O: Oxygen is always −II in compounds (except peroxides, superoxides and with fluorine) so the hydrogens in water must each be +I.

$Mg(OH)_2$: Magnesium is +II, oxygen −II and hydrogen +I – each element has its usual oxidation number in compounds.

H_2: Here hydrogen is an uncombined element so its oxidation number is zero.

Looking at the oxidation numbers in the equation above, the two hydride ions in magnesium hydride have been oxidised from −I to 0 (and two of the hydrogens in water have been reduced from +I to 0).

The half-equation is

$$2H^- \rightarrow H_2 + 2e^-$$

which could also be written

$$2H^- - 2e^- \rightarrow H_2$$

(d)   The fuel tank of one type of hydrogen-powered car holds 70 kg of MgH_2.

What *volume* of hydrogen gas, measured at room temperature and pressure, would be released if this amount of magnesium hydride reacted with water?

(Assume 1 mole of molecules of a gas at room temperature and pressure has a volume of 24 dm^3.)

[Relative atomic masses: H = 1, Mg = 24]

The relative molecular mass of magnesium hydride is $24 + (2 \times 1) = 26$.

70 kg is 70 000 g which is $\dfrac{70\,000}{26} = 2692.3077$ mol.

The equation tells us that 1 mol of magnesium hydride produces 2 mol of hydrogen so the volume will be

$2692.3077 \times 2 \times 24 = 129\,230.77$ dm^3

Volume of hydrogen $= 1.3 \times 10^5$ dm^3

WHY DOES IT MATTER?

Alternative fuels

Cars run almost exclusively on fuels derived from crude oil – petrol and diesel. These are non-renewable and will eventually run out. Electric cars also run largely on fossil fuels: the electricity is generated at power stations, most of which burn coal or oil.

Several renewable fuels have been suggested including hydrogen, mentioned in the question. Liquid fuels are the most convenient as they do not escape like gases and they can be pumped and stored using existing technology.

Ethanol can be made by fermentation of sugar and other carbohydrate crops and can be used as a fuel (mixed with petrol as 'gasohol' in some countries). Its disadvantage is that it produces rather less energy per gram than petrol. Another possibility is so-called rape methyl ester, a fuel made from rapeseed oil which can be used to replace diesel. Rape is the yellow crop which is becoming common in the UK. Its oil is a triglyceride (an ester of glycerol and three molecules of fatty acid). Rape methyl ester is made by hydrolysing the oil to produce the fatty acids – see Fig. 4.1b – and converting them to their methyl esters which are of similar volatility to diesel fuel.

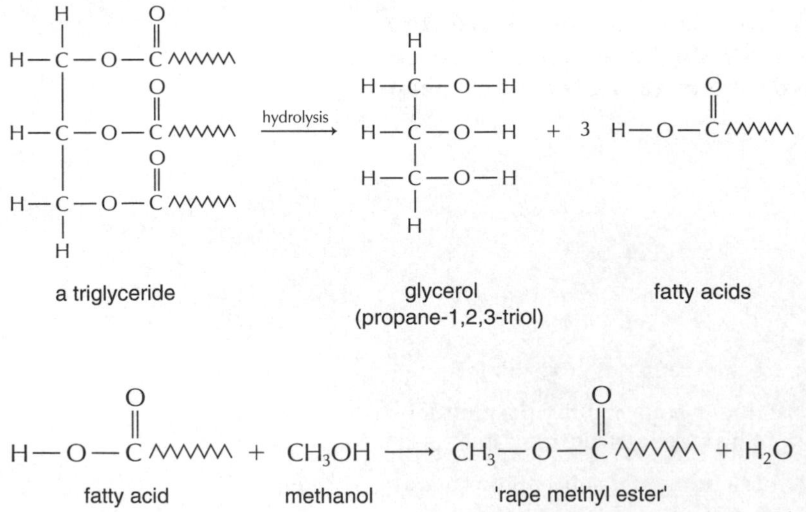

a triglyceride   glycerol (propane-1,2,3-triol)   fatty acids

fatty acid   methanol   'rape methyl ester'

Fig. 4.1b

4.2   In petrol-driven cars, a mixture of hydrocarbons and air is ignited with a spark. The exhaust fumes from such cars contain various atmospheric pollutants such as NO, NO_2, CO and hydrocarbons. The pollutants can be removed from the exhaust fumes by use of a *catalytic converter* made from an aluminium alloy in the form of a fine mesh, coated with platinum or rhodium. The catalytic converter promotes reactions between molecules of the pollutant gases which result in less harmful exhaust fumes.

(a) (i) How do you account for the presence of carbon monoxide in exhaust fumes?

Carbon monoxide results from the incomplete combustion of the hydrocarbon fuel.

Complete combustion of a hydrocarbon would lead to the formation of carbon dioxide and water.

(ii)   Write equations for the formation of the two oxides of nitrogen.

$$N_2(g) + O_2(g) \rightarrow 2NO(g)$$

$$N_2(g) + 2O_2(g) \rightarrow 2NO_2(g)$$

Don't forget to make sure that the equations are balanced (same number of atoms of both elements on each side of the arrow) and that you have included state symbols.

(iii) Suggest a reason why reactions involving nitrogen gas require a high temperature.

The nitrogen molecule has a strong N≡N triple bond. To bring about reaction, this must be broken, a process requiring a lot of energy.

The N≡N bond energy is $945 \, kJ \, mol^{-1}$, one of the strongest covalent bonds. Compare it with that of a typical covalent bond, C—C, which is $347 \, kJ \, mol^{-1}$.

(b) (i) Explain how a catalyst affects the rate of a chemical reaction.

Catalysts speed up reactions by allowing them to take place via a different pathway which has a lower activation energy.

Activation energy is the energy barrier between the reactants and the transition state (or activated complex) of a reaction – see Fig. 4.2a. It represents the energy which has to be put in to start breaking bonds in the reactants. The strong N≡N bond means that reactions involving N_2 molecules tend to have high activation energies.

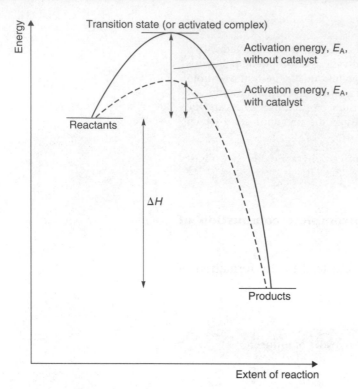

Fig. 4.2a The reaction profile for an exothermic reaction.

(ii) Why is the catalyst in the form of a fine mesh?

The fine mesh gives the catalyst a large surface area on which the reactions can take place.

The catalyst in a catalytic converter is a so-called **heterogeneous catalyst**, that is the catalyst is in a different phase from the reactants and products. In this case the catalyst is solid and the reactants and products are gases. In heterogeneous catalysis, molecules of the reactants are adsorbed on (form weak chemical bonds with) the surface of the catalyst. This bonding can either weaken the bonds in the reactants making it easier for them to react or hold the reacting molecules in just the right positions to react together or both.

Homogeneous catalysts are in the same phase as the reactants, for example the acid catalyst for the esterification reaction between an alcohol and a carboxylic acid where all the species are in aqueous solution.

(iii) Why do you think that manufacturers recommend that cars fitted with catalytic converters should only use unleaded petrol?

Unleaded petrol must be used as lead compounds 'poison' the catalyst surface by being adsorbed on to it and blocking the surface to the gaseous reactants – a common problem with heterogeneous catalysts.

(iv) Explain why a high temperature is necessary for the catalytic converter to operate.

High temperatures speed up reactions (by supplying the activation energy required). In a converter the reactions must take place quickly as the gases are swept rapidly over the catalyst.

Typically, exhaust fumes spend only about 1/10 second in the converter, so reactions must take place rapidly and converters work effectively only when they are hot (over 300 °C). It is possible that motorists who do many short stop–start journeys never get their converters to the effective temperature.

(c) Suggest a method by which the concentrations of CO and hydrocarbons in exhaust fumes could both be reduced, other than by using catalysts.

One method is to ensure that there is sufficient air mixed with the fuel to allow complete combustion of the fuel to water and carbon dioxide.

As always, 'suggest' means that there is no one correct answer and that credit can be obtained for any one of a variety of sensible approaches.

WHY DOES IT MATTER?

In a catalytic converter, reactions take place such as

$$2CO(g) + 2NO(g) \rightarrow 2CO_2(g) + N_2(g)$$

and

hydrocarbons + nitrogen oxides $\rightarrow$ carbon dioxide + nitrogen

+ water

The effects of the various pollutants found in car exhaust fumes are as follows.

Carbon monoxide is toxic as it combines more effectively with haemoglobin in the blood than does oxygen. It thus reduces the blood's oxygen-carrying capacity so that victims can die of suffocation even when there is plenty of oxygen in the air. Carbon monoxide poisoning is a common form of death, both accidental and deliberate.

Nitrogen oxides are toxic. They form acids in moist air thus contributing to acid rain. They also combine with hydrocarbons in sunlight to form a photochemical smog containing ozone which makes breathing difficult.

Unburnt hydrocarbons help form photochemical smogs (see above).

Carbon dioxide, which is emitted from converters, is not wholly innocuous. It is a greenhouse gas which may be contributing to global warming.

It is interesting to calculate the ratio of fuel to air which would just produce complete combustion. If we assume petrol is octane, C_8H_{18}, of relative molecular mass 114, the balanced equation is

$$C_8H_{18}(g) + 12\tfrac{1}{2}O_2(g) \rightarrow 8CO_2(g) + 9H_2O(g)$$

But each mole of oxygen has four moles of nitrogen mixed with it in air. So 1 mol (114 g) of octane requires 12.5 mol of O_2 and 50 mol of N_2. This is 12.5×32 g (= 400 g) of O_2 + 50×28 g (=1400 g) of N_2.

This is a total of 1800 g of air to 114 g of octane, a ratio of 15.8:1 by mass.

4.3 In 1930 the American Thomas Midgeley inhaled a lungful of the CFC with formula CCl_2F_2 and used it to blow out a candle. He did this to demonstrate a new refrigerant which was non-flammable and non-toxic.

(a) What does CFC stand for?

C̲hlorof̲luoroc̲arbon.

(b) State **two** other uses (as well as refrigerants) to which CFCs have been put.

Aerosol propellants, 'blowing' of plastic foams such as expanded polystyrene (used for packaging), cleaning solvents (for grease), fire extinguishers.

(c) In the stratosphere, CFCs are broken down to give chlorine atoms. These chlorine atoms react with ozone molecules thus:

$$Cl^{\bullet} + O_3 \rightarrow ClO^{\bullet} + O_2 \qquad [1]$$

(i) Write equation [2] which shows how the $ClO^{\bullet}$ radicals can react further with oxygen atoms to give oxygen molecules and chlorine atoms.

$ClO^{\bullet} + O \rightarrow O_2 + Cl^{\bullet}$

Note that there is one species with an unpaired electron (a radical, sometimes called a free radical) on the left of the equation, so there must also be one on the right. Make sure that your equation is balanced; two O atoms and one Cl on the left, the same number of each on the right.

(ii) Use equations [1] and [2] to give an overall equation. Then explain why a few chlorine atoms can cause the destruction of many ozone molecules.

The overall equation is

$O_3 + O \rightarrow 2O_2$

Ozone (trioxygen) molecules are converted into oxygen (dioxygen). Chlorine atoms are converted into ClO· radicals and back to chlorine atoms and so the process can be repeated many times. They are regenerated, not used up, so they effectively catalyse the reaction.

$ClO^{\bullet}$ radicals could also be regarded as the catalyst.

To get the overall equation, add together each side of the two equations, cancelling the species which appear on both sides.

$$Cl^{\bullet} + O_3 \rightarrow ClO^{\bullet} + O_2$$
$$ClO^{\bullet} + O \rightarrow O_2 + Cl^{\bullet}$$
$$\overline{O_3 + O + \cancel{Cl^{\bullet}} + \cancel{ClO^{\bullet}} \rightarrow 2O_2 + \cancel{Cl^{\bullet}} + \cancel{ClO^{\bullet}}}$$

(iii) The reaction represented by equation [1] is *first order* with respect to both chlorine atoms and ozone molecules. Write the *rate equation* for this reaction.

Rate = $k\,[Cl^{\bullet}]\,[O_3]$

Remember the use of [] to represent concentration in $mol\,dm^{-3}$.

A reaction is first order with respect to a reactant if the rate is directly proportional to the concentration of that reactant (i.e. the concentration raised to the power of one). So this reaction is first order with respect to both $Cl^{\cdot}$ and O_3 (and so, incidentally, is second order overall).

k is the **rate constant** for the reaction.

(iv) What happens to the rate of ozone depletion by chlorine atoms if the concentration of chlorine atoms is doubled? Explain your answer.

The rate will double because the reaction is first order with respect to chlorine atoms.

This is what first order means: double the concentration, double the rate.

(v) Give the equation for a *termination* reaction in which chlorine atoms are removed from the stratosphere.

$Cl^{\cdot} + Cl^{\cdot} \rightarrow Cl_2$

Termination occurs when two radicals combine together to produce a species with no unpaired electrons.

$$\overset{\times\,\times}{\underset{\times\,\times}{\times}Cl}\times \quad + \quad \overset{\circ\,\circ}{\underset{\circ\,\circ}{}}Cl\overset{\circ}{\underset{\circ}{}} \quad \rightarrow \quad \overset{\times\,\times}{\underset{\times\,\times}{\times}}Cl\overset{\circ\,\circ}{\underset{\circ\,\circ}{\circ}}Cl\circ$$

(vi) Why is the rate of this termination reaction slow in the stratosphere?

The concentration of $Cl^{\cdot}$ in the atmosphere is very small, so two $Cl^{\cdot}$s are unlikely to collide often.

(d) If the production of CFCs were to be phased out now, it would be a long time before their damaging effects on the ozone layer began to be reduced. Suggest *two* reasons for this.

• **Each $Cl^{\cdot}$ can destroy many O_3 molecules before a termination occurs.**
• **There is a large reservoir of CFCs 'locked up' in insulation foams, fridges and unused aerosol cans which will slowly be released as the products are scrapped or used.**
• **CFCs in the lower atmosphere will take a long time to reach the stratosphere where they are broken down into chlorine atoms.**

Any two of the above points would give full marks.

(e) Suggest a reason why Midgeley was unaware of the drawbacks associated with CFCs.

It is unlikely that in Midgeley's time much research had been done on the reactions of CFCs, especially UV light-induced reactions with ozone.

or

The role of ozone in absorbing UV light was probably less well understood at this time.

NOTE

'Suggest' means that there are several possible answers each of which will gain marks provided that it is sensible and relevant. The word 'suggest' often indicates that you have to use your chemistry knowledge to think about the question rather than know the right answer.

This question is even more speculative than the last one. Many answers are likely to gain marks provided that they are sensibly thought out and explained. Your suggestions do not need to be lengthy or detailed.

> (f) Chemists are now looking for compounds to replace CFCs. Describe **two** properties which these compounds should have, as well as being non-toxic and non-flammable.

No smell, similar volatility to existing CFCs so that they can be used in existing equipment without modification, stable with respect to breakdown in the stratosphere.

Any two of the above would gain credit. Other sensible suggestions are possible.

> (g) A variety of halogenoalkanes similar to Midgeley's original compound have been synthesised to see whether they might be useful. For example, CF_2I_2 has been synthesised.
> (i) Explain why this compound has a higher boiling point than CCl_2F_2.

The compound has a higher relative molecular mass (and therefore more electrons) than CCl_2F_2 and therefore the van der Waals forces between the molecules will be greater.

Van der Waals forces depend on the number of electrons in the species. This, of course, means that they increase as relative molecular mass increases. An answer based on the fact that CF_2I_2 will have a greater permanent dipole than CCl_2F_2 would also gain credit.

> (ii) Give a reason why CF_2I_2 reacts faster than CCl_2F_2 with sodium hydroxide solution.

The C—I bond is weaker than (has a lower energy than) the C—Cl bond.

The reaction is

$$CF_2I_2 + OH^- \rightarrow CF_2IOH + I^-$$

so the C—I bond must break.

The shared electrons in C—I are further from the I nucleus than the shared electrons in C—F are from the F nucleus, so the C—I bond is weaker.

> (iii) When CF_2I_2 reacts with sodium hydroxide, iodide ions are formed in the alkaline solution. In order to test for the presence of these ions a suitable substance has first to be added to neutralise the alkali.
>
> If you were doing this test, state:
> the substance you would use to carry out the neutralisation;
> the substance you would add to test for the iodide ions;
> the positive result of the test.

This is a question about the use of silver nitrate to test for halide ions (Cl^-, Br^-, I^-). Don't be put off by the unfamiliar compound CF_2I_2.

Nitric acid.

CARE

Do not compare the strength of the C—I bond with the C—F bond, which is so strong that it is unlikely to break.

CARE

The C—Cl bond is more polar than the C—I bond, leading to a greater $\delta+$ charge on the C atom, making it more susceptible to nucleophilic attack. Experiment shows that bond strength is more important than bond polarity in this reaction.

Hydrochloric acid would produce chloride ions which would also react with the silver nitrate which is used to test for the iodide ions. Sulphuric acid would produce sulphate ions which would also react with silver nitrate.

Silver nitrate solution.

A yellow precipitate (of silver iodide).

This is a test for all halide ions – formation of a precipitate with silver nitrate solution – white for chloride, cream for bromide and yellow for iodide.

$$Ag^+(aq) + I^-(aq) \rightarrow AgI(s)$$

To distinguish more clearly between iodide and bromide, add aqueous ammonia: the precipitate will dissolve in the case of bromide but not in the case of iodide.

WHY DOES IT MATTER?

CFCs had a great variety of uses until the 1980s when scientists began to understand the part they were playing in the destruction of the ozone layer.

CFCs are also 'greenhouse gases' in the atmosphere, that is they allow light energy from the Sun to pass through but absorb heat energy which is radiated from the Earth's surface. This is thought to be contributing to global warming. Because of the environmental problems that they cause, a declaration was signed in 1990 by over sixty nations agreeing to phase out CFCs before the end of the century. Alternatives are being developed including HCFCs – hydrochlorofluorocarbons such as CF_3CHCl_2.

A number of CFCs has been used including $CFCl_3$, CF_2Cl_2, $CF_2ClCFCl_2$. The first has the systematic name trichlorofluoromethane. Try to name the other two. (The answers are below.)

Incidentally, Thomas Midgeley was responsible for another environmental problem. It was he who introduced the idea of adding lead compounds to petrol to make it burn more smoothly in high compression engines. Midgeley, of course, had no idea of the problems his ideas would cause – he was simply unlucky!

Answers

1,1,2-Trichloro-2,2-trifluoroethane

Dichlorodifluoromethane

5 BIOCHEMISTRY

What you should know now

❑ The meaning of the Michaelis constant.

❑ Enzymes are protein-based catalysts.

❑ An enzyme has an active site into which the substrate molecule fits.

❑ Enzymes' shapes can be altered by relatively small changes in pH or temperature.

❑ DNA and RNA are molecules involved in passing on hereditary information.

❑ How to write balanced symbol equations for combustion reactions.

❑ Understand the ways in which monosaccharides link together to form polysaccharides.

❑ The pathway by which glucose is used to supply energy in cells.

❑ The formulae of amino acids.

❑ Formation of peptides from amino acids.

❑ The structures of proteins.

Exam questions

Q 5.1 (a) The enzyme hexokinase catalyses the conversion of glucose into glucose-6-phosphate with a Michaelis constant (K_M) for glucose of $100\,\mu\text{mol}\,l^{-1}$ and an optimum pH of 7.5. The graph below shows the formation of glucose-6-phosphate when the enzyme acts on glucose at $37\,°\text{C}$ and pH 7.5 with an initial glucose concentration of $50\,\mu\text{mol}\,l^{-1}$ in the reaction mixture.

(i) Calculate the initial rate of the reaction.

(ii) State the effect on the initial rate of decreasing the pH to 6.0. Explain why this change occurs.

(iii) Give **three** reasons why the concentration–time graph is not linear and, in each case, explain your answer.   9 marks

(b) DNA and RNA are both nucleic acids found in cells.

(i) Give **one** structural and **two** chemical differences between DNA and RNA.

(ii) Name **three** different types of RNA. State the function of each type in the cell.   6 marks

[N, '94]

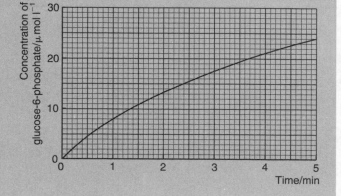

5.2 (a) (i) Write a balanced equation for the complete oxidation of glucose.

(ii) In what form is the energy released when glucose undergoes combustion?

(iii) In cells much of the energy released is stored in a chemical form. Name the compound used to store the energy.

(iv) Explain briefly why this compound can act as an energy store in the cell.

(v) Give **two** distinct uses of this compound in the cell, stating in each case what form of energy results. 　7 marks

(b) (i) Name the **two** main storage molecules found in humans.

(ii) Explain the need for these two types of storage molecule. 　4 marks

(c) Cellulose is a major structural macromolecule found in plants. Draw a section of the cellulose molecule showing two of the repeating units. Name the molecule which forms the repeating units and name the type of bond which links them. 　4 marks

[N, '94]

5.3 (a) The amino acid, 2-aminopropanoic acid (*alanine*), has the formula $CH_3CH(NH_2)COOH$.

(i) Write down the structural formulae of the predominant ionic species present at pH 1, pH 7 and pH 13.

(ii) Give the formula of the peptide, alanylalanylalanine. 　4 marks

(b) Many proteins show considerable amounts of secondary structure.

(i) Name **two** types of secondary structure commonly found.

(ii) Draw a diagram to show the essential features of **one** of these structures.

(iii) How are these structures stabilised? 　5 marks

[N, '96 (specimen)]

Answers

5.1 (a) The enzyme hexokinase catalyses the conversion of glucose into glucose-6-phosphate with a Michaelis constant (K_M) for glucose of $100\,\mu mol\,l^{-1}$ and an optimum pH of 7.5. The graph below shows the formation of glucose-6-phosphate when the enzyme acts on glucose at 37 °C and pH 7.5 with an initial glucose concentration of $50\,\mu mol\,l^{-1}$ in the reaction mixture.

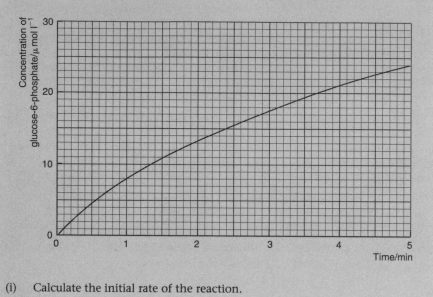

(i) Calculate the initial rate of the reaction.

The initial rate is given by the gradient (slope) of the tangent to the graph at time 0. Here the tangent goes through the point $27\,\mu mol\,dm^{-3}$ at 3 min (see Fig. 5.1a) so the gradient (and therefore the initial rate) is

$27\,\mu mol\,dm^{-3}/3\,min = 9\,\mu mol\,dm^{-3}\,min^{-1}$

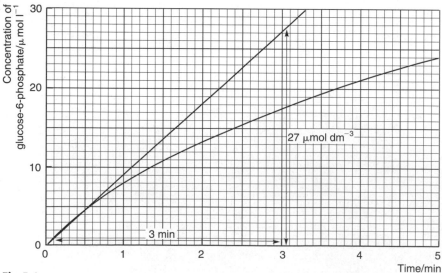

Fig. 5.1a

Since it is not possible to draw any tangent exactly, you would be allowed some leeway in your answer but make sure you show working on the graph.

The rate of any chemical reaction (enzyme-catalysed or not) is given by the slope or gradient of the concentration–time graph. To find the rate at any given time, draw the tangent to the graph at that point and work out its gradient. To do this, pick a pair of convenient points on the tangent and divide the vertical height difference by the horizontal displacement.

In this case we are measuring the initial rate so we draw the tangent at time 0. Initial rates are important in reaction kinetics as time 0 is the only point when we know the concentrations of all the species in the reaction mixture *exactly* (because we have just measured them out).

> (ii)   State the effect on the initial rate of decreasing the pH to 6.0. Explain why this change occurs.

Effect:

This would decrease the rate.

You don't need to know anything about the reaction to answer this; remember the question states that 7.5 is the *optimum* pH, so any change of pH, up or down, must decrease the rate.

Reason:

Changing the pH will affect the shape of the enzyme and therefore reduce its catalytic effect.

Enzymes have an active site into which their substrate fits (the 'lock and key' theory). If the shape of this active site changes they will not be as effective as catalysts. Increasing acidity (decreasing pH) may protonate the lone pairs on oxygen atoms in the CONH groups of the enzyme's protein chain and affect their ability to form hydrogen bonds with other parts of the chain. It is these bonds which help to establish the tertiary structure of the protein and govern its shape – see Fig. 5.1b.

> (iii)   Give **three** reasons why the concentration–time graph is not linear and, in each case, explain your answer.

Reason 1:

The concentration of the substrate will decrease as the reaction proceeds.

Explanation: Since the concentration of the substrate is less than the Michaelis constant, the rate will be proportional to the substrate concentration.

At low concentrations of substrate, enzyme-catalysed reactions are first order with respect to the substrate concentration (i.e. the rate is directly proportional to the substrate concentration). So the reduction of the concentration of the substrate as it is used up in the reaction reduces the rate. At higher concentrations of substrate, the enzyme becomes saturated with substrate. So increasing the substrate concentration no longer affects the rate (we say the reaction is zero order with respect to the substrate) and the reaction attains a maximum rate, V_{max}. The idea of enzyme-catalysed reactions changing from first to zero order as the enzyme becomes saturated with substrate is called Michaelis–Menten kinetics. The Michaelis constant (K_M) is the concentration at which the rate is $\frac{1}{2}V_{max}$. This reaction has a Michaelis constant for glucose of $100\,\mu mol\,dm^{-3}$, so at the initial concentration specified ($50\,\mu mol\,dm^{-3}$) the glucose concentration will still affect the rate.

CARE

When measuring the vertical height difference and the horizontal displacement you must use the scale of the graph, not just count squares.

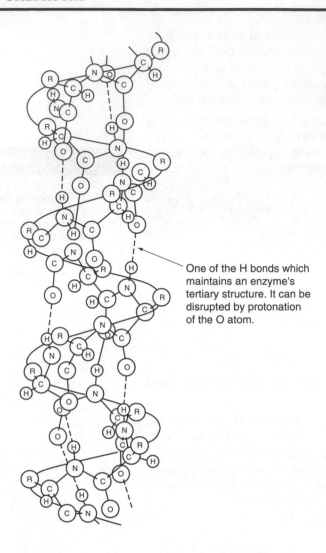

One of the H bonds which maintains an enzyme's tertiary structure. It can be disrupted by protonation of the O atom.

Fig. 5.1b

Reason 2:

As the reaction proceeds, the concentration of the product (glucose-6-phosphate) increases.

Explanation:

The product inhibits the enzyme. This is so that when a cell has a high concentration of glucose-6-phosphate and needs no more, no more will be produced.

Reason 3:

Enzymes will denature over a period of time.

Explanation:

Over time, all enzymes will denature (change their shape) and become less active.

(b) DNA and RNA are both nucleic acids found in cells.
(i) Give **one** structural and **two** chemical differences between DNA and RNA.

Structural difference:

DNA has a double helix structure and RNA a single helix.

Chemical difference 1:

DNA contains deoxyribose units whilst RNA contains ribose units.

Chemical difference 2:

DNA contains the base thymine while RNA contains uracil.

DNA (deoxyribonucleic acid) and RNA (ribonucleic acid) are both polymers of monomers called nucleotides. Nucleotides consist of a phosphate bonded to a ribose sugar and a base – see Fig. 5.1c.

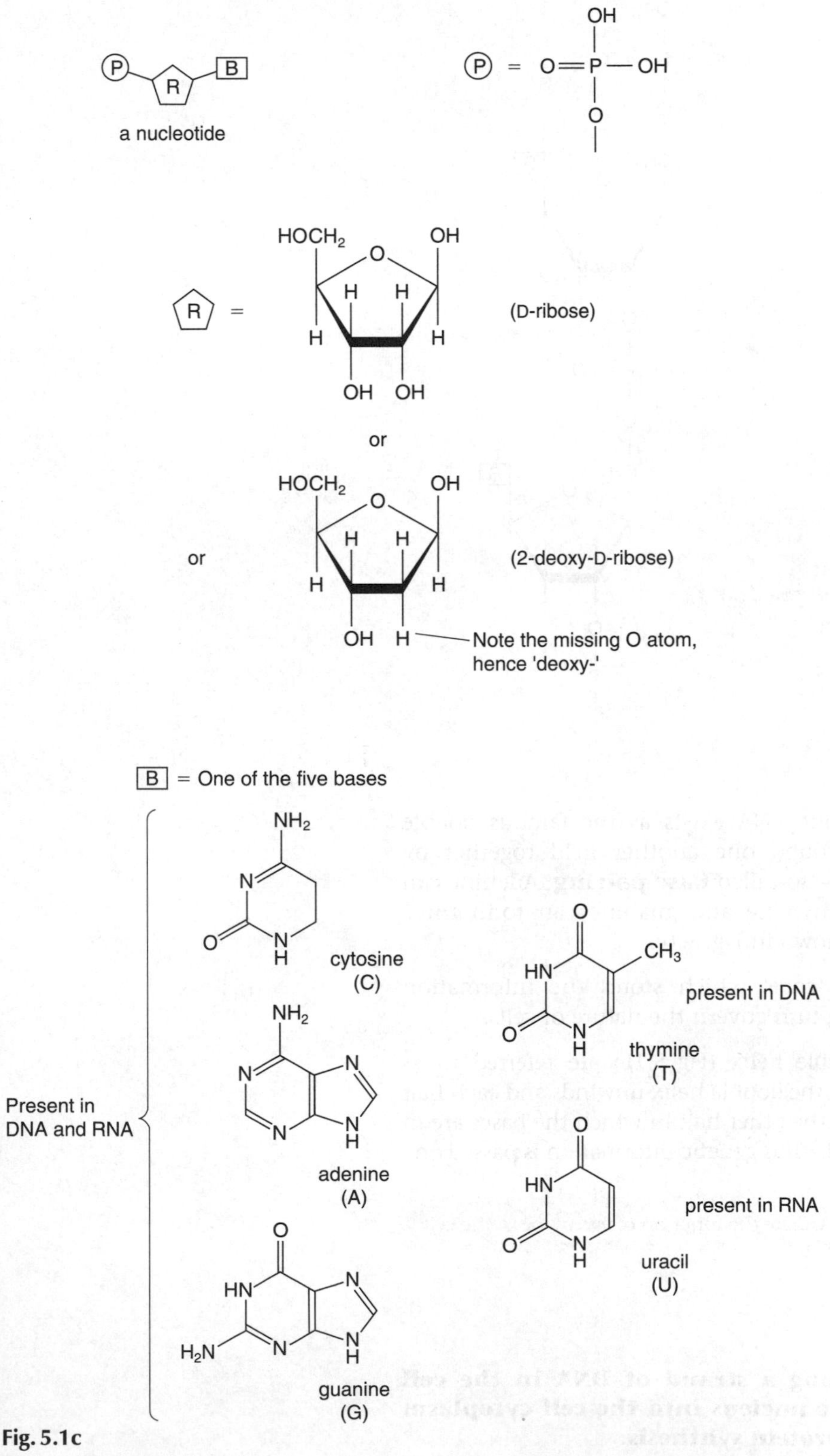

Fig. 5.1c

The nucleotides polymerise as shown in Fig. 5.1d by elimination of water from a phosphate group on one nucleotide and an —OH group on the next.

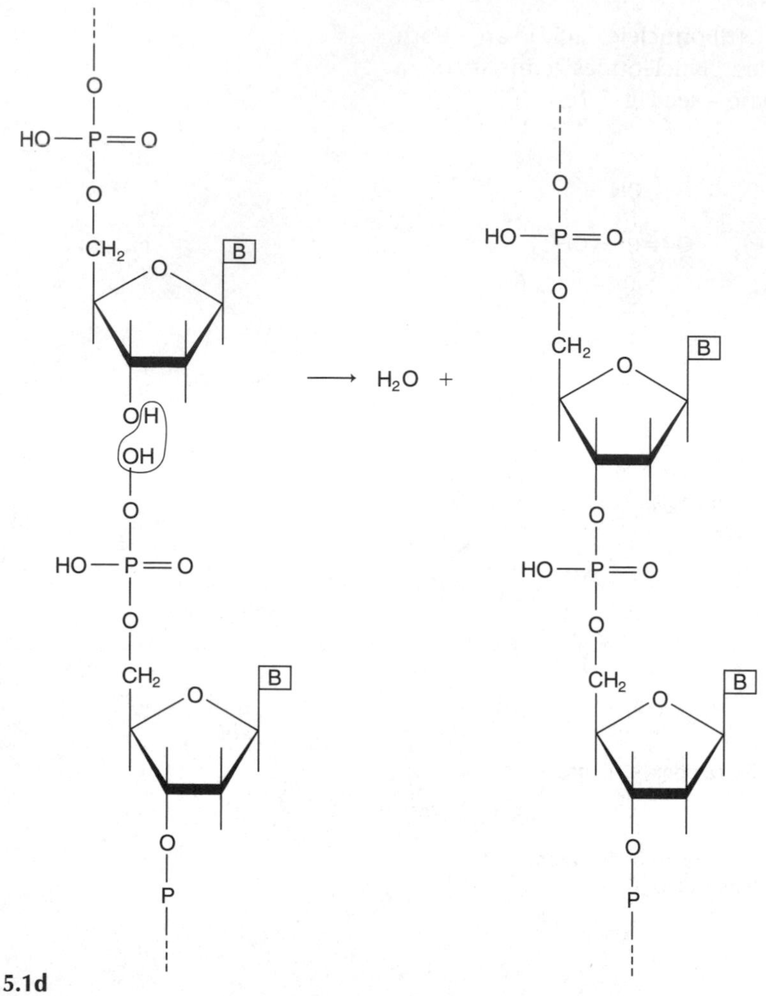

Fig. 5.1d

RNA is generally single-stranded but DNA exists as the famous double helix where two strands wind around one another held together by hydrogen bonds between the bases – so-called **base pairing**. Adenine can form two hydrogen bonds with thymine and guanine can form three hydrogen bonds with cytosine as shown in Fig. 5.1e.

The order of the bases forms a 'code' which stores the information required to make proteins, which in turn govern the nature of cells.

The two strands of the DNA double helix (Fig. 5.1f) are referred to as complementary. Durling replication, the double helix unwinds and each half acts as a template for a new copy of the other half in which the bases are in exactly the same order as the original. Thus genetic information is passed on.

(ii)   Name **three** different types of RNA. State the function of each type in the cell.

RNA 1:

Messenger RNA (mRNA)

Function: This RNA forms along a strand of DNA in the cell nucleus and migrates from the nucleus into the cell cytoplasm carrying the information for protein synthesis.

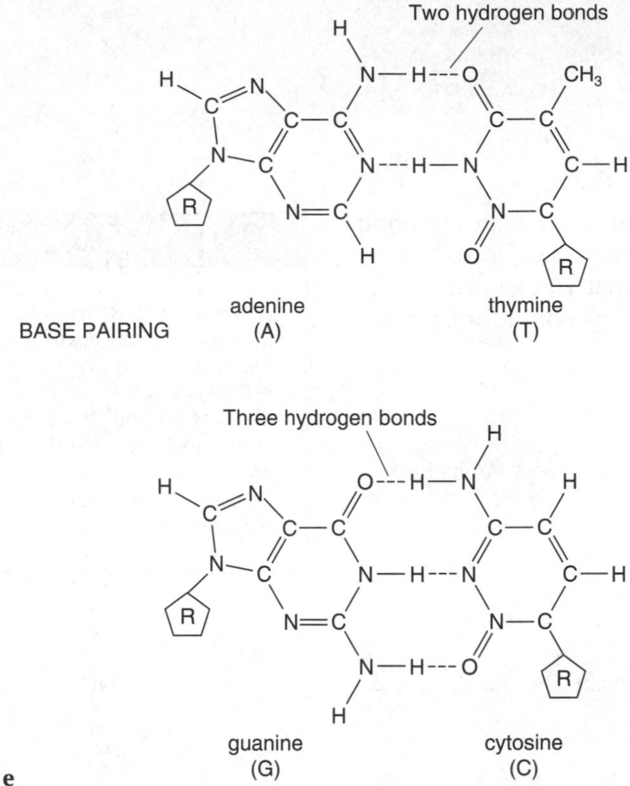

BASE PAIRING

adenine (A)   thymine (T)

Two hydrogen bonds

Three hydrogen bonds

guanine (G)   cytosine (C)

Fig. 5.1e

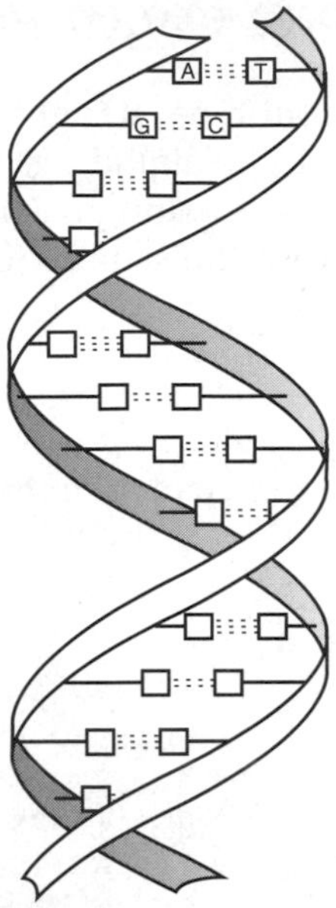

Fig. 5.1f

RNA 2:

Transfer RNA (tRNA)

Function: This RNA binds at one end to the mRNA and at the other to amino acids and so 'reads' the genetic code and assembles protein from separate amino acids in a sequence determined by that of the mRNA which obtained it from the DNA in the nucleus.

RNA 3:

Ribosomal RNA (rRNA)

Function: This RNA is found in the ribosome. It helps to bind together the enzymes in the ribosome which catalyse the buildup of protein from individual amino acids in the sequence provided by the mRNA.

WHY DOES IT MATTER?

Genetic fingerprinting

Each person has different DNA from that of every other individual (except an identical twin). This DNA is found in every cell in the body. In genetic fingerprinting, the DNA is extracted and cut into short fragments by enzymes. These fragments, which are electrically charged, are separated by electrophoresis. This involves pulling the fragments through a gel using an electric field. The smallest and most highly charged fragments move fastest. After treatment to make them visible, a pattern looking rather like a bar code is produced. This can be used to compare biological material, such as blood or semen, found at the scene of a crime with DNA from the suspect.

5.2 (a) (i) Write a balanced equation for the complete oxidation of glucose.

$$C_6H_{12}O_6(s) + 6O_2(g) \rightarrow 6CO_2(g) + 6H_2O(l)$$

This should be straightforward. It is worth putting in state symbols (s = solid, l = liquid, g = gas, aq = aqueous solution, i.e. dissolved in water) when writing balanced equations to ensure that you get full marks. In this case the symbol for water could be l or g as the conditions are not specified.

EXAM TIP

Always check that your balanced equations really are balanced by adding up the numbers of atoms of each element on both sides of the arrow and checking that they are the same: in this case, 6 carbons, 12 hydrogens and 18 oxygens.

(ii) In what form is the energy released when glucose undergoes combustion?

Heat.

(iii) In cells much of the energy released is stored in a chemical form. Name the compound used to store the energy.

Adenosine triphosphate, ATP.

(iv) Explain briefly why this compound can act as an energy store in the cell.

When ATP is formed from adenosine diphosphate (ADP), energy has to be put in (i.e. the reaction is endothermic). This means that the products have more energy than the reactants. So reversing this reaction (ATP→ADP) gives out energy.

In fact ΔH for the ADP → ATP reaction is $+32\,\text{kJ}\,\text{mol}^{-1}$.

In the cell, direct oxidation of glucose as in the answer to (i) would produce so much heat that the cell would be destroyed. In fact glucose is oxidised gradually in a series of steps, which involve the conversion of ADP to ATP:

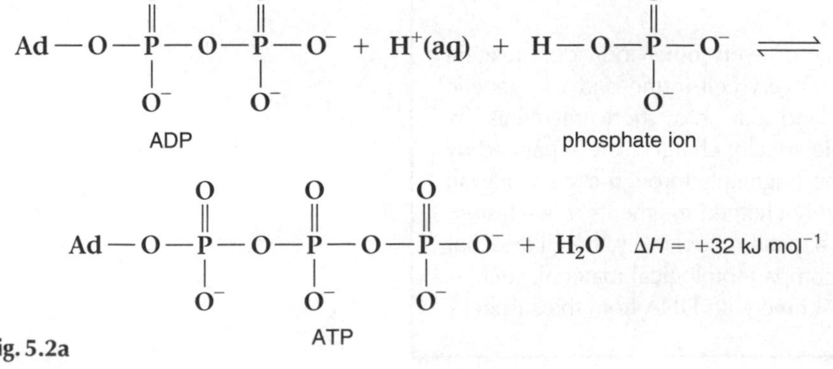

Fig. 5.2a

where Ad represents adenosine:

Ad =

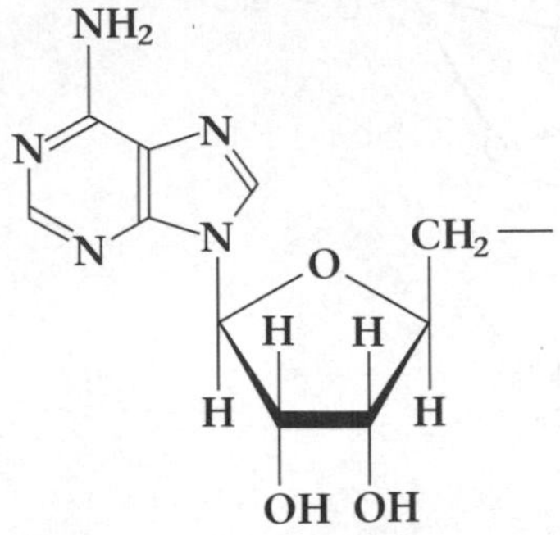

Fig. 5.2b

(v)   Give **two** distinct uses of this compound in the cell, stating in each case what form of energy results.

Use 1: Muscle contraction.

Form of energy: mechanical.

This form of energy could be expressed in different ways, e.g. gravitational potential energy if the muscle contraction is used to lift something or kinetic energy if it is used to move something.

Use 2: Synthesis of new chemical compounds.

Form of energy: chemical.

Other answers could include: use: active transport of molecules through the cell membrane; form of energy: potential.

(b) (i) Name the **two** main storage molecules found in humans.

Molecule 1: glycogen.

Molecule 2: fats (triglycerides).

(ii)   Explain the need for these two types of storage molecule.

Glycogen is used as a short-term energy store in the cell as it is slightly soluble and therefore more available.

Fats are used for longer-term storage as they are compact and insoluble.

These two molecules represent part of a hierarchy of energy storage:

ATP   glucose   glycogen   fats   protein

←————————————— shorter term
                 longer term —————————→

(c)   Cellulose is a major structural macromolecule found in plants. Draw a section of the cellulose molecule showing two of the repeating units. Name the molecule which forms the repeating units and name the type of bond which links them.

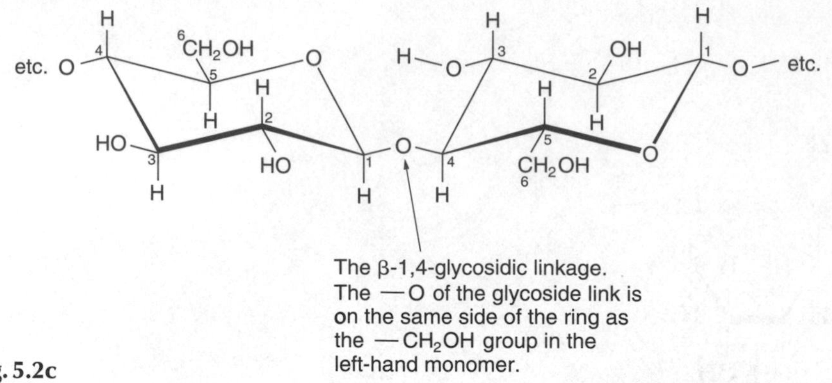

The β-1,4-glycosidic linkage.
The —O of the glycoside link is
on the same side of the ring as
the — CH$_2$OH group in the
left-hand monomer.

Fig. 5.2c

The repeating units or monomers are molecules of β-glucose and the type of bond which links them is a β-1,4-glycosidic linkage.

Glucose can exist in α- and β-forms (a pair of stereoisomers) which are shown in Fig. 5.2d with the conventional numbering of the rings.

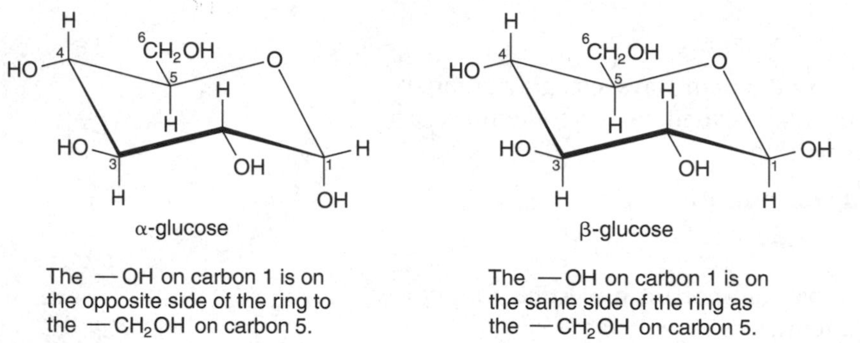

α-glucose

The —OH on carbon 1 is on
the opposite side of the ring to
the —CH$_2$OH on carbon 5.

β-glucose

The —OH on carbon 1 is on
the same side of the ring as
the —CH$_2$OH on carbon 5.

Fig. 5.2d

In the α-form, the —OH group on carbon number 1 is on the opposite side of the ring to the —CH$_2$OH group on carbon number 5, i.e. one sticks up above the plane of the ring and the other down below it. In the β-form, both the —OH group at carbon 1 and the —CH$_2$OH at carbon 5 are on the same side of the ring.

The C—O—C link between two sugars in cellulose and other polysaccharides is called a glycosidic linkage. The β tells us the shape of the glucose molecules involved in the bond and the -1,4- which carbons in the two glucose units are involved.

WHY DOES IT MATTER?

Polysaccharides

Single sugars such as glucose are called monosaccharides. They can join together with the loss of a water molecule from two —OH groups (one on each sugar) to form a glycosidic link and thus form polysaccharides such as cellulose.

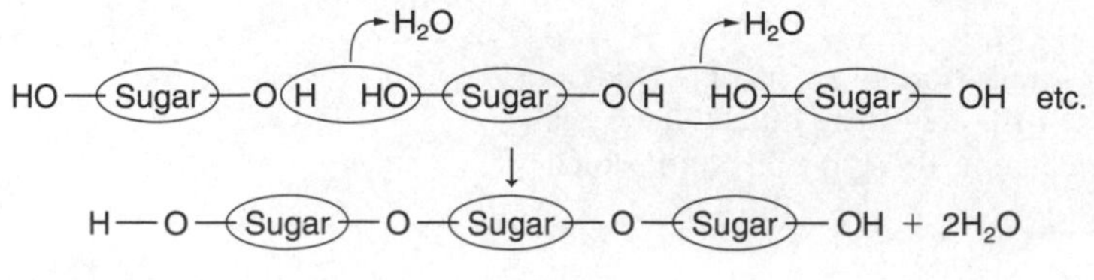

A large variety of polysaccharides are possible depending on

- the type of sugar which polymerises, e.g. glucose or fructose
- the isomer of a particular sugar which polymerises, e.g. α- or β-glucose
- the carbons which link, e.g. a -1,4- or a -1,6- link
- the degree of branching and cross-linking between the chains.

These differences bring about large differences in properties.

We have seen that cellulose is a polyglucose with a β-1,4-glycosidic linkage. The shape of the glycosidic linkages in the cellulose molecule allows the chains to lie parallel to one another and to form hydrogen bonds between them. This accounts for the strength of cellulose which is a structural material in plants and is the compound from which cotton is largely made.

Starch, the polysaccharide in foods such as bread and potatoes, has two forms – α-amylose and amylopectin. Both have α-1,4-glycosidic linkages but α-amylose is linear and amylopectin branched. Humans have enzymes which can digest starch but not cellulose.

Glycogen, the food storage polysaccharide found in animal cells, is similar to amylopectin but has more branching through α-1,6-linkages.

In each case, the polymer is insoluble but individual glucose units are soluble. In the organism, the insoluble polymers are hydrolysed with the aid of enzymes to soluble monomers which can be transported around the organism to where they are needed to be oxidised to release energy.

5.3 (a) The amino acid, 2-aminopropanoic acid (*alanine*), has the formula $CH_3CH(NH_2)COOH$.
(i) Write down the structural formulae of the predominant ionic species present at pH 1, pH 7 and pH 13.

pH 1:

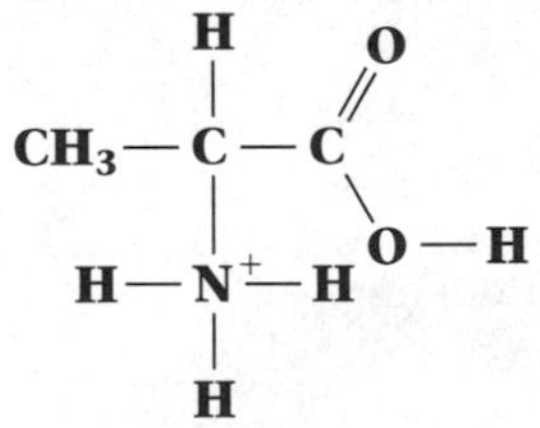

pH 1 is acidic and so the lone pair of the —NH$_2$ group will be protonated.

pH 7:

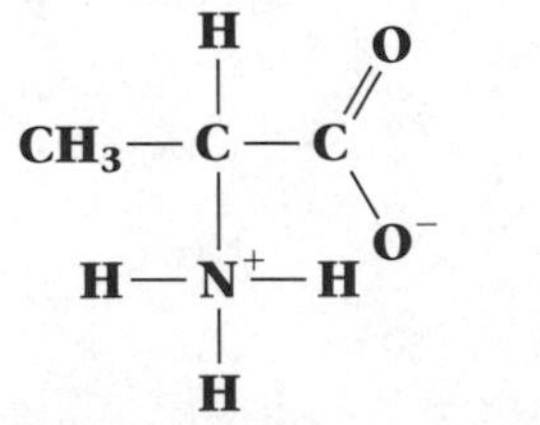

pH 7 is neutral, so the —NH$_2$ group will be protonated and the —CO$_2$H group will be deprotonated to —CO$_2^-$. This species with a positive and a negative charge but which is neutral overall is called a zwitterion.

pH 13:

$$CH_3-\underset{\underset{\underset{H}{|}}{\underset{|}{N}}}{\overset{\overset{H}{|}}{C}}-\overset{\overset{O}{\|}}{C}\overset{O^-}{}$$

pH 13 is strongly basic. The —NH₂ group will be deprotonated as will the —CO₂H, existing as —CO₂⁻.

(ii)   Give the formula of the peptide, alanylalanylalanine.

The tripeptide's structure is

$$H_2N-\underset{\underset{CH_3}{|}}{\overset{\overset{H}{|}}{C}}-\overset{\overset{O}{\|}}{C}-\underset{\underset{H}{|}}{N}-\underset{\underset{CH_3}{|}}{\overset{\overset{H}{|}}{C}}-\overset{\overset{O}{\|}}{C}-\underset{\underset{H}{|}}{N}-\underset{\underset{CH_3}{|}}{\overset{\overset{H}{|}}{C}}-\overset{\overset{O}{\|}}{\underset{O-H}{C}}$$

This tripeptide (one containing three amino acids) has three alanine molecules joined together via peptide bonds. These are formed by loss of water from an —NH₂ group and a —CO₂H group as shown.

$$\overset{\overset{O}{\|}}{-C}\underset{}{\overset{\frown H_2O}{-\boxed{O-H\quad H}-}}\underset{}{N-}\quad\longrightarrow\quad\overset{\overset{O}{\|}}{-C}-\underset{\underset{H}{|}}{N}-$$

the peptide or
amide bond

(b)   Many proteins show considerable amounts of secondary structure.
(i)   Name **two** types of secondary structure commonly found.

The secondary structure of a protein is the folding of the backbone of the protein's polypeptide chain.

Structure 1 is the α-helix in which the polypeptide chains are wound into a spiral.

Structure 2 is the β-pleated sheet or just β-sheet.

(ii)   Draw a diagram to show the essential features of **one** of these structures.

α-helix

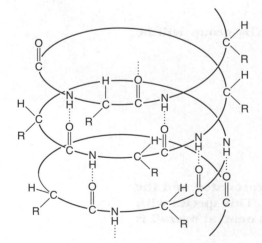

Fig. 5.3a

β-sheet

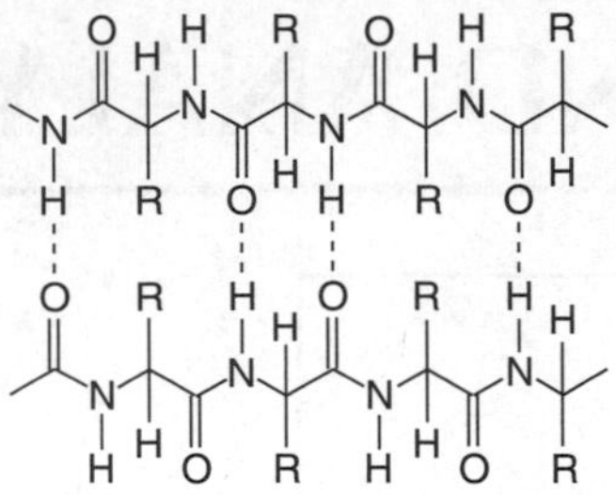

Fig. 5.3b

You need draw just one.

(iii) How are these structures stabilised?

Both structures are held in position by hydrogen bonding between C=O groups and NH groups in the peptide groups of the chain.

The **primary structure** of a protein is the sequence of amino acids along the chain.

The **secondary structure** describes the regular arrangement of sections of the chain as described above.

Tertiary structure describes further coiling or folding rather as a coiled telephone wire might tangle. This level of structure is held together by hydrogen bonding, other intermolecular forces and also by sulphur–sulphur bridges which link two amino acids called cystine to form a molecule of cysteine.

Quaternary structure describes the way in which large protein molecules are made up of more than one polypeptide chain and possibly other non-polypeptide components. Haemoglobin, for example, has four polypeptide units and four haem groups containing iron ions.

WHY DOES IT MATTER?

Why can't we 'unboil' an egg?

Hydrogen bonds are about 20 times weaker than covalent bonds. The primary structure of a protein is held together by covalent amide (peptide) bonds.

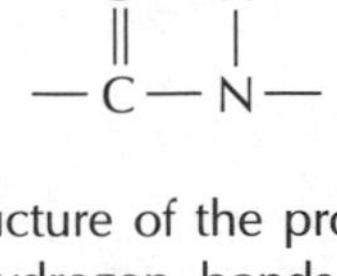

the amide (peptide) bond

The secondary structure of the protein determines the shape of a protein and is held by hydrogen bonds. Heating an egg to 100 °C for three minutes is enough to break many of the hydrogen bonds and disrupt the secondary structure of the protein albumin in egg white. It is not sufficient to break covalent bonds, so the primary structure remains unaltered. However, as we cool the boiled egg, it is enormously unlikely that the hydrogen bonds will re-form in exactly the same way as they were before heating. So the protein is irreversibly changed and boiled eggs do not 'unboil'.

6 SPECTROSCOPY AND ANALYSIS

What you should know now

❏ Mass spectrometers form ions from their samples and separate them according to their mass:charge ratios, by accelerating them and deflecting them in magnetic and electric fields.

❏ Infra-red spectroscopy (IR) measures the absorption of infra-red radiation by the sample, which occurs when parts of the sample molecule vibrate.

❏ In order to absorb infra-red radiation and therefore appear in an IR spectrum, a bond must have a dipole and its vibration must produce change in dipole moment.

❏ Proton nuclear magnetic resonance spectroscopy (proton NMR) measures the energy absorbed by protons as they 'flip' in a magnetic field. It tells us about the environment of hydrogen atoms in the sample molecule.

❏ Low-resolution mass spectrometry measures relative molecular masses to the nearest whole number. High-resolution mass spectrometry can measure relative masses to five or six decimal places.

❏ At high resolution, the peaks in a NMR spectrum are split because the applied magnetic field is affected by the magnetic fields of other nuclei in the molecule.

❏ Ultra-violet/visible spectroscopy measures the absorption of ultra-violet and visible radiation which promotes electrons from one orbital to another.

❏ X-ray diffraction enables the positions of atoms to be found by looking at the way in which X-rays are scattered by substances.

❏ Isomers are compounds with the same molecular formula but a different arrangement of the atoms in space.

❏ The empirical formula of a compound is the simplest whole number ratio of the atoms present.

❏ The molecular formula of a compound is the number of each type of atom in its molecule. It is a multiple of the empirical formula.

Exam questions

6.1 (a) (i) Draw a simple diagram of a mass spectrometer, labelling clearly the major components of the system. Draw the paths through the apparatus of a heavy ion and a light ion at a given intensity of the magnetic field.
(ii) In which region of the instrument does ion fragmentation take place?
(iii) What can be deduced about the composition of a compound if it shows two molecular ion peaks of equal intensity? 9 marks

(b) A compound is known to have either structure **A** or **B**, and is investigated by infra-red spectroscopy and proton magnetic resonance spectroscopy.

CH_3CH_2—⬡—OH

A

CH_3—⬡—O—CH_3

B

(i) The infra-red spectrum of the compound measured in solution in trichloromethane shows a peak at a wavenumber of $3600\ cm^{-1}$, the appearance of which is strongly dependent on the concentration of the solution. With this information, assign structure **A** or **B** to the compound and give an appropriate explanation.
(ii) The proton magnetic resonance spectrum shows (with other peaks) a triplet peak and a quartet peak system. State which groups of protons in the compound are responsible for these two multiplets.
(iii) Explain how deuterium oxide could be used to distinguish between structures **A** and **B** by proton magnetic resonance spectroscopy. 6 marks

[N, '94]

6.2 A sample of mineral water from a plastic bottle was tested for contamination by compound **C**.

⬡ with substituents:
$$\overset{O}{\overset{\|}{C}}—O—CH_2CH_3$$
$$\overset{}{\underset{O}{\overset{\|}{C}}}—O—CH_2CH_3$$

C

(a) In order to identify **C** by infra-red spectroscopy, it was first extracted from the water and concentrated to give an oily liquid.
(i) Give **one** reason why the substance was extracted from the water before the infra-red spectrum was measured.
(ii) If a sample of pure **C** were available, how would infra-red spectroscopy be used to confirm the structure of the extracted material?
(iii) The extracted oil contains residual water, which shows three well-defined absorption peaks in the infra-red spectrum. Draw simple diagrams to show the types of vibration which give rise to the three absorption peaks.
(iv) What is the characteristic of a vibrational mode in a water molecule which ensures that infra-red radiation can be absorbed? 5 marks

(b) The mass spectrum of the extract in (a) shows a molecular ion with $m/e = 222$.
(i) Explain briefly how ions are produced in the mass spectrometer.
(ii) Explain briefly how the resulting ions are separated and analysed in the mass spectrometer.
(iii) The molecular ion of $m/e = 222$ corresponds not only to the molecular formula of **C** but also to other formulae (e.g. $C_{13}H_{18}O_3$). Explain how measurement of the value of m/e may be used to distinguish between such molecular formulae. 5 marks

(c) The proton magnetic resonance spectrum of **C** is measured in solution in deuterotrichloromethane, using tetramethylsilane as an internal reference.
(i) Why does tetramethylsilane show a single resonance peak?

(ii) The ethyl group protons in **C** appear as characteristic quartet and triplet signals in the spectrum. Explain with the aid of a simple diagram why the CH_2 protons in the ethyl group appear as a quartet.

5 marks

[N, '93]

Q6.3 Two organic compounds, **A** and **B** are isomers with the composition by mass of carbon, 70.5%; hydrogen, 5.9%; oxygen, 23.6%. **A** is moderately soluble in water and **B** is a pleasant-smelling liquid. Their mass spectra are shown below.

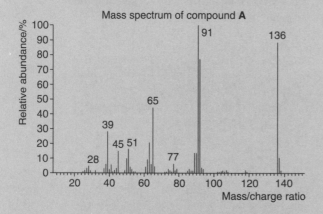

Mass spectrum of compound **A**

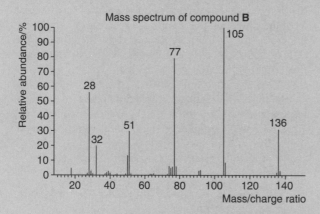

Mass spectrum of compound **B**

(a) (i) What is the empirical formula of **A** and **B**? (Relative atomic masses: C = 12, O = 16, H = 1)
(ii) What is the molecular formula of **A** and **B**? Justify your answer.

4 marks

(b) Give the formulae of the molecular fragments corresponding to the following peaks: Mass/charge ratio: 136, 105, 91, 77.

4 marks

(c) What structural formulae would you predict for **A** and **B**?

3 marks

(d) Describe **two** tests or chemical reactions in which the behaviour of **A** and **B** would differ.

2 marks

[L (Nuffield)]

Q6.4 (a) Consider the following instrumental techniques used in structural analysis.

infra-red spectroscopy
ultra-violet/visible spectroscopy
proton magnetic resonance spectroscopy
mass spectrometry
X-ray diffraction

(i) Which of these techniques uses electromagnetic radiation of the lowest frequency?
(ii) Which technique uses electromagnetic radiation of the shortest wavelength?
(iii) In which technique is electromagnetic radiation absorbed by vibrating bonds?
(iv) Which technique is the most generally appropriate for examining isotopically labelled molecules?
(v) Which technique can be used to determine the concentration of a compound in solution down to levels of approximately $10^{-5}\,mol\,l^{-1}$?
(vi) Indicate which technique often makes use of deuterated solvents and explain why this is so.

7 marks

(b) The visible absorption spectrum of the water effluent from a textile dyeworks is shown below. The effluent contains two dyes **X** and **Y**, whose absorption bands are shown.

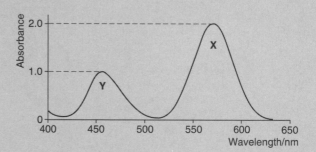

(i) Explain the process whereby a molecule absorbs energy from light.
(ii) Define the *absorbance* (A) of a solution in terms of the incident light intensity (I_0) and the emergent light intensity (I).
(iii) State the expression which relates the molar extinction (absorption) coefficient (ε) of a substance to the absorbance of its solution at λ_{max}, the molar concentration (c) and the pathlength (l) of the solution.
(iv) Deduce from the above spectrum the *relative* concentrations of **X** and **Y** in the effluent, given that the ε_{max} value for **X** is three times that of **Y**.
(v) On a particular occasion it was found that, in the absence of **X** and **Y**, the effluent was brown in colour. State why the absorbance measurements would give erroneous concentration values for **X** and **Y** when these two compounds are present in such an effluent. How might the error be minimised if a sample of the effluent containing no **X** or **Y** were available?

8 marks

[N, '94]

Answers

6.1 (a) (i) Draw a simple diagram of a mass spectrometer, labelling clearly the major components of the system. Draw the paths through the apparatus of a heavy ion and a light ion at a given intensity of the magnetic field.

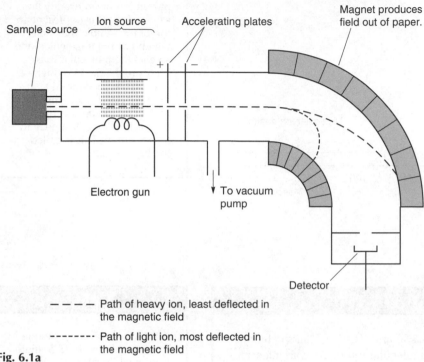

– – – – Path of heavy ion, least deflected in
the magnetic field

- - - - - - - Path of light ion, most deflected in
the magnetic field

Fig. 6.1a

Note the 'hockey stick' shape of the mass spectrometer. Heavy ions have more momentum than light ones and are less deflected in the magnetic field.

(ii) In which region of the instrument does ion fragmentation take place?

Fragmentation takes place in the ion source after formation of the ions.

The bombardment of the sample molecules by high-energy electrons gives the resulting ions sufficient energy to break one or more bonds.

(iii) What can be deduced about the composition of a compound if it shows two molecular ion peaks of equal intensity?

This suggests that the sample contains bromine.

This suggests that one element in the sample molecule exists as a pair of isotopes of approximately equal abundance. A possible element is bromine, which consists of ^{79}Br (50.5%) and ^{81}Br (49.5%). A bromine-containing compound would give rise to two molecular ion peaks of virtually equal intensity separated by two mass units.

Fig. 6.1b is the mass spectrum of 2-bromopropane, showing the two parent ion peaks.

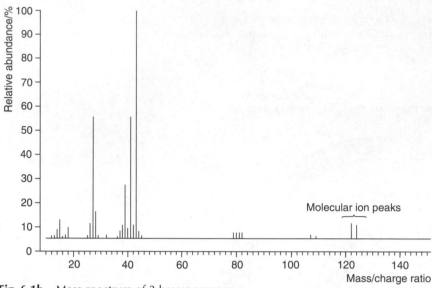

Fig. 6.1b   Mass spectrum of 2-bromopropane.

EXAM TIP

The terms **molecular ion** and **parent ion** mean exactly the same thing. They represent an ion of the sample molecule which has not fragmented and so give the molecular mass. They are the peaks of highest mass in the spectrum (i.e. at the right-hand side). The molecular ion peak is not necessarily the tallest peak in the spectrum. This is called the **base peak**.

BY THE WAY

Chlorine exists as the isotopes ^{35}Cl (75.5%) and ^{37}Cl (24.5%), so chlorine-containing compounds will show two molecular ion peaks with intensities in the ratio approximately 3:1 with the smaller peak two mass units greater than the larger one.

Fig. 6.1c is the mass spectrum of chloroethane showing the two parent ion peaks at mass 64 and 66 in the approximate abundance ratio 3:1. Can you suggest what the two peaks at masses 49 and 51 (also in the ratio 3:1) are?

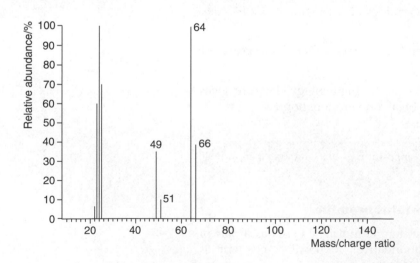

Fig. 6.1c   Mass spectrum of chloroethane.

CARE

Although the average atomic mass of chlorine is 35.5. there is no peak in the mass spectrum of chlorine at mass 35.5: ions of each isotope pass individually through the mass spectrometer and are not averaged.

BY THE WAY

There may be small peaks in the mass spectrum at higher mass (to the right) of the parent ion peak. At first sight this seems impossible! These peaks are caused by molecules containing isotopes. For example about 1% of all carbon atoms are ^{13}C. So a molecule containing one carbon atom would have an isotope peak at one mass unit above that of the parent ion but only 1/100 as abundant. A molecule containing two carbon atoms has twice the chance of containing at least one atom of ^{13}C and so has a peak at one mass unit above the molecular ion of abundance 2/100 that of the molecular ion. A three-carbon molecule will have an '$M + 1$' peak of abundance 3/100 that of the parent ion and so on. This can be a useful way of working out how many carbon atoms a molecule contains.

(b)   A compound is known to have either structure **A** or **B**, and is investigated by infra-red spectroscopy and proton magnetic resonance spectroscopy.

(i)   The infra-red spectrum of the compound measured in solution in trichloromethane shows a peak at a wavenumber of 3600 cm^{-1}, the appearance of which is strongly dependent on the concentration of the solution. With this information, assign structure **A** or **B** to the compound and give an appropriate explanation.

The compound has structure A.

A peak which varies in appearance with concentration suggests intermolecular hydrogen bonding, which is possible in A but not B.

The peak at 3600 cm^{-1} is the O—H stretch of the alcohol. Alcohols can hydrogen bond together as shown in Fig. 6.1d. This effectively increases the mass of the H atom and therefore lowers the frequency of the vibration of the O—H bond. In a more concentrated solution, more hydrogen bonding is likely as the molecules of **A** are more likely to approach one another. Thus the band's appearance depends on concentration.

Fig. 6.1d

EXAM TIP

It is useful to learn a few IR frequencies. That at about 1700 cm^{-1} is a C=O stretch, that at about 3300 cm^{-1} is an O—H stretch, although the exact frequency varies as in the main text.

(ii) The proton magnetic resonance spectrum shows (with other peaks) a triplet peak and a quartet peak system. State which groups of protons in the compound are responsible for these two multiplets.

The CH_3 protons of A produce the triplet peak.

They are split by coupling with the CH_2 protons which can line up in three ways, affecting the magnetic field felt by the CH_2 protons.

The CH_2 protons of A give rise to the quartet.

They couple with the CH_3 protons which can line up in four ways. See 6.2 for more detail.

(iii) Explain how deuterium oxide could be used to distinguish between structures **A** and **B** by proton magnetic resonance spectroscopy.

The OH hydrogen in A is 'labile', that is it will exchange with hydrogen atoms from a solvent such as water. If deuterium oxide (D_2O) is used as a solvent, the OH hydrogen will exchange with deuterium:

$$ROH + D_2O \rightleftharpoons ROD + HDO$$

The NMR peak due to the OH hydrogen will disappear as deuterium does not give an NMR spectrum.

B has no labile hydrogens and so its spectrum will be unaffected by D_2O.

WHY DOES IT MATTER?

Magnetic resonance imaging

Protons will give different NMR signals depending on their precise environments. The body contains protons in water, fats, proteins and many other compounds. Different organs in the body contain different proportions of these compounds and can therefore be differentiated by NMR. A tumour, for example, would give a different signal to that from healthy tissue. To do a scan, the body is passed through a large doughnut-shaped magnet and irradiated with radiofrequency radiation. NMR has two advantages over X-rays:

- radio waves have less energy than X-rays (at the intensity used) and therefore do not cause tissue damage

- X-rays are best at detecting heavy atoms (which are relatively uncommon in the body except in bones) while NMR looks at protons which are found all over the body

When used in medicine, NMR is called magnetic resonance imaging to avoid the word 'nuclear' which some patients associate with radioactivity.

6.2 A sample of mineral water from a plastic bottle was tested for contamination by compound **C**.

$$\text{benzene ring} - \underset{\underset{O}{\|}}{C} - O - CH_2CH_3$$
$$\text{benzene ring} - \underset{\underset{O}{\|}}{C} - O - CH_2CH_3$$

C

(a)   In order to identify **C** by infra-red spectroscopy, it was first extracted from the water and concentrated to give an oily liquid.

(i)   Give **one** reason why the substance was extracted from the water before the infra-red spectrum was measured.

As a contaminant, C would be present in only a small concentration in the water sample. It would therefore absorb very little infra-red and give very small peaks in the spectrum.

Or, putting this another way, water itself absorbs IR and might swamp the spectrum of **C**.

(ii)   If a sample of pure **C** were available, how would infra-red spectroscopy be used to confirm the structure of the extracted material?

The IR spectrum of a sample of pure C could be matched with that of the extract from the mineral water.

A close match, especially in the fingerprint region, between $1400\,\text{cm}^{-1}$ and $900\,\text{cm}^{-1}$, would indicate that the two were identical. This region of the spectrum often contains complex vibrations of the whole molecule and is different for different compounds.

(iii)   The extracted oil contains residual water, which shows three well-defined absorption peaks in the infra-red spectrum. Draw simple diagrams to show the types of vibration which give rise to the three absorption peaks.

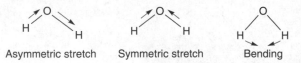

Fig. 6.2a   Asymmetric stretch   Symmetric stretch   Bending

(iv)   What is the characteristic of a vibrational mode in a water molecule which ensures that infra-red radiation can be absorbed?

In order to absorb infra-red radiation, a molecular vibration in a water molecule (or any other molecule) must involve a change in dipole moment as the molecule vibrates.

BY THE WAY

A non-linear molecule has $3n - 6$ vibrational modes and a linear one $3n - 5$, where n is the number of atoms in the molecule. So water, which has three atoms and is non-linear, has $(3 \times 3) - 6 = 3$.

All the vibrations of a water molecule involve a change in dipole moment as shown in Fig. 6.2b.

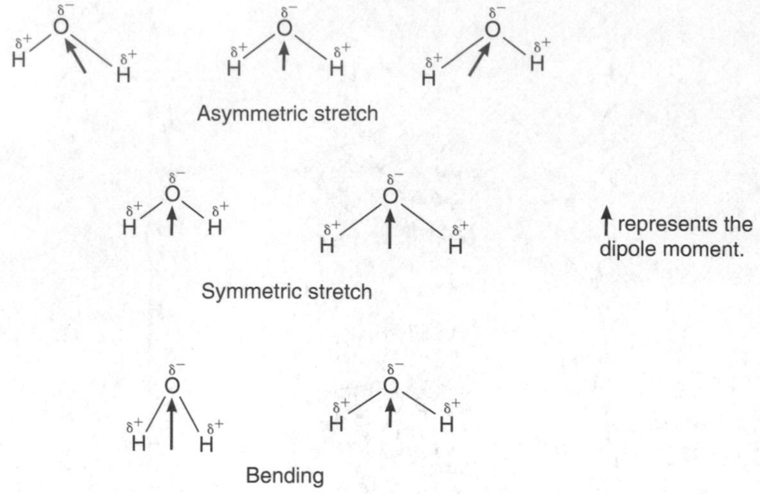

Asymmetric stretch

Symmetric stretch

↑ represents the dipole moment.

Bending

Fig. 6.2b

A vibration such as the symmetrical stretching vibration in carbon dioxide does not involve a change in dipole moment as the dipoles of the two bonds cancel out – Fig. 6.2c.

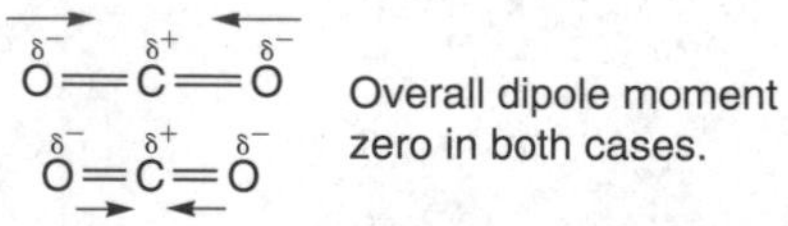

Overall dipole moment zero in both cases.

Fig. 6.2c

(b)   The mass spectrum of the extract in (a) shows a molecular ion with $m/e = 222$.
(i)   Explain briefly how ions are produced in the mass spectrometer.

The sample is vaporised in a vacuum and then bombarded with a stream of electrons from an electron gun. These knock out one (or sometimes more) electrons from the sample molecules leaving positive ions. If S is the sample molecule:

$$S + e^- \rightarrow S^+ + 2e^-$$

There are other methods of ion formation – see 6.3. Once formed, the ions may fragment – see 6.1 and 6.3.

(ii)   Explain briefly how the resulting ions are separated and analysed in the mass spectrometer.

The positive ions are accelerated towards a negatively charged plate and some pass through a hole and form a beam. This beam is then deflected into a circular path by a magnetic field so that it hits a detector. Heavier ions need a stronger magnetic field to deflect them into the detector and so the masses of the sample molecule and any fragments can be found.

(iii) The molecular ion of $m/e = 222$ corresponds not only to the molecular formula of **C** but also to other formulae (e.g. $C_{13}H_{18}O_3$). Explain how measurement of the value of m/e may be used to distinguish between such molecular formulae.

To distinguish between ions of the same whole number mass but different molecular formulae requires high resolution mass spectrometry. **Here the masses are measured to, say, five places of decimals. By using accurate values of relative atomic masses, the accurate values of relative molecular masses of different empirical formulae can be worked out and compared with the data from the mass spectrum.**

For example, the molecular formula of C is $C_{12}H_{14}O_4$ so it has a relative molecular mass of 222 using $C = 12$, $H = 1$ and $O = 16$. However, if we use accurate values for the relative atomic masses ($C = 12.00000$, $H = 1.00782$, $O = 15.99492$), its relative molecular mass is 222.08916. A compound with molecular formula $C_{13}H_{18}O_3$ would have an accurate relative molecular mass of 222.12552. These can be distinguished easily by a high resolution mass spectrum. Check the figures for yourself!

(c) The proton magnetic resonance spectrum of **C** is measured in solution in deuterotrichloromethane, using tetramethylsilane as an internal reference.
(i) Why does tetramethylsilane show a single resonance peak?

All its protons are in an identical environment.

Tetramethylsilane is

BY THE WAY

Deuterotrichloromethane, which has a deuterium isotope of hydrogen (^2H) in place of a normal hydrogen atom (^1H), is used as a solvent because deuterium nuclei have no spin. So there is no NMR signal from the solvent to affect the spectrum of the sample.

(ii) The ethyl group protons in **C** appear as characteristic quartet and triplet signals in the spectrum. Explain with the aid of a simple diagram why the CH_2 protons in the ethyl group appear as a quartet.

This is an example of spin–spin splitting. Protons themselves produce a magnetic field. The CH_2 protons in an ethyl group (CH_3CH_2—) will experience the magnetic field applied by the NMR spectrometer modified by the nearby protons of the CH_3 group. These will add to or cancel part of the applied field depending on whether each CH_3 proton lines up with or against the applied field. There are four possibilities illustrated in Fig. 6.2d. The CH_2 peak is therefore split into a quartet of peaks with areas in the ratio $1:3:3:1$.

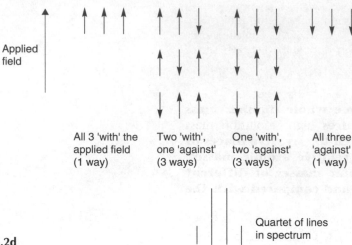

Fig. 6.2d

The CH_3 protons in an ethyl group will experience the magnetic field applied by the NMR spectrometer modified by the nearby protons of the CH_2 group. Both protons can line up with the applied field, strengthening it slightly, both can line up against it, weakening it slightly and one can line up with and one against it, cancelling out their effects. This last possibility can come about in two ways (one proton with the applied field, the other proton against and vice versa) and so is twice as likely as either of the others – see Fig. 6.2e.

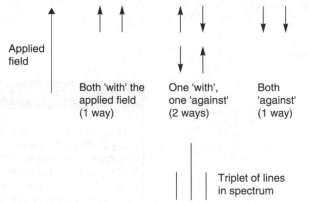

Fig. 6.2e

This has the effect of splitting the CH_3 peak into three smaller peaks (a triplet) with areas in the ratio $1:2:1$.

The whole spectrum of the ethyl group would look like Fig. 6.2f.

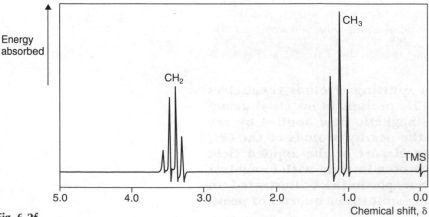

Fig. 6.2f

BY THE WAY

Mass and mass to charge ratio (*m/e*)

Strictly speaking mass spectrometers do not measure mass but mass to charge ratio (*m/e*), where *m* is the relative molecular mass and *e* the number of positive charges on the ion. During ionisation by electron impact, most molecules will lose just one electron to form an ion with one positive charge. However, some might be ionised twice to form ions with two positive charges. These will be deflected twice as much in the mass spectrometer and will in fact behave like a singly charged ion of half the mass. So a singly charged ion of relative mass 50 would appear at the same place in the spectrum as a doubly charged ion of mass 100. So the horizontal axes of mass spectra are labelled *m/e* rather than *m*. In practice there is seldom a problem as double ionisation is fairly rare.

WHY DOES IT MATTER?

The Intoximeter

If you are suspected of drunken driving, a police officer will first do a roadside breath test. This uses a fuel cell to estimate the concentration of alcohol (ethanol) in your breath but it is not accurate enough to be used in evidence. At the police station, a suspect may be asked to give a further breath sample to be tested on an 'Intoximeter'. This measures the absorption of IR radiation at $2950\,cm^{-1}$ – the frequency at which the C—H bonds in ethanol absorb. A beam of IR is passed alternately through a 28 cm long cell containing the breath sample and an identical empty cell and a detector compares the intensities. The more alcohol in the sample, the more IR will be absorbed and the less will reach the detector.

Diabetics might have propanone, which also has C—H bonds, in their breath. The intoximeter has a second sensor which allows for this.

Printer to produce hard copy of the test results

Keyboard to enter details of the suspect and operator

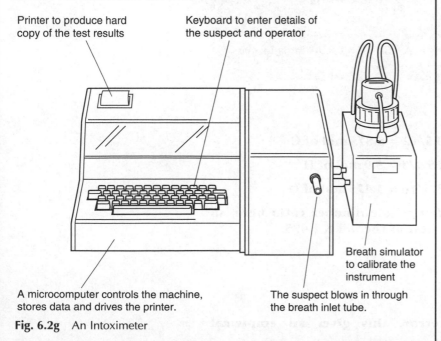

Breath simulator to calibrate the instrument

A microcomputer controls the machine, stores data and drives the printer.

The suspect blows in through the breath inlet tube.

Fig. 6.2g An Intoximeter

6.3 Two organic compounds, **A** and **B**, are isomers with the composition by mass of carbon, 70.5%; hydrogen, 5.9%; oxygen, 23.6%. **A** is moderately soluble in water and **B** is a pleasant-smelling liquid. Their mass spectra are shown below.

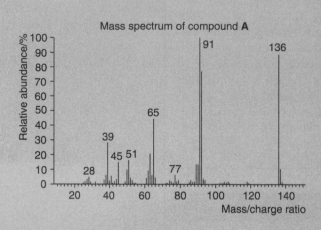

Mass spectrum of compound **A**

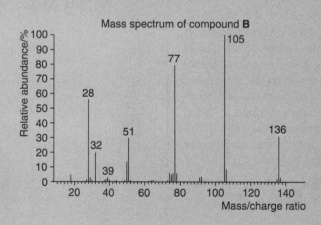

Mass spectrum of compound **B**

(a) (i) What is the empirical formula of **A** and **B**? (Relative atomic masses: C = 12, O = 16, H = 1)

In 100 g of compound:

the no. of moles of carbon is 70.5/12 = 5.875 mol of C

the no. of moles of hydrogen is 5.9/1 = 5.9 mol of H

the no. of moles of oxygen is 23.6/16 = 1.475 mol of O

It is not easy to see the simplest whole number ratio here, so divide each number by the smallest of them, i.e. 1.475:

C: 5.875/1.475 = 3.98

H: 5.9/1.475 = 4.0

O: 1.475/1.475 = 1.0

Allowing for experimental error, this gives an empirical formula of C_4H_4O.

To find the empirical formula, we first need to know the number of moles of each element in the compound. The question gives the percentage composition of the two isomers (which are, of course, the same); these figures are the same as the number of grams of each element in 100 g of compound.

> (ii) What is the molecular formula of **A** and **B**? Justify your answer.

The relative mass of the empirical formula unit C_4H_4O would be

$$(12 \times 4) + (1 \times 4) + (16 \times 1) = 68$$

From the mass spectrum, the actual relative molecular mass is twice this (136), so the molecular formula must be twice the empirical formula: $C_8H_8O_2$.

To convert the empirical formula to a molecular formula, we need to know the relative molecular mass. This is found from the peak of highest mass to charge ratio in the mass spectra. This is 136. Ignore the tiny peaks just above this: they are caused by ions containing isotopes such as ^{13}C which have higher masses.

> (b) Give the formulae of the molecular fragments corresponding to the following peaks: mass/charge ratio: 136, 105, 91, 77.

136: this is the parent ion, formula $C_8H_8O_2$.

105: formula C_7H_5O

To produce an ion of relative mass 105 (136 − 31), a fragment of mass 31 must have broken off the parent ion. A fragment of this mass must contain either one carbon and one oxygen or two carbons.

$C + O = 12 + 16 = 28$ and would therefore need three Hs to give a mass of 31, i.e. a fragment of formula CH_3O. Loss of this leaves a fragment of mass 105.

$C + C = 12 + 12 = 24$ and would therefore need 7 Hs to give a mass of 31. It is not possible for such a fragment to occur in bonding terms.

91: formula C_7H_7

To produce an ion of relative mass 91 (136 − 45), a fragment of mass 45 must have broken off the parent ion. A fragment of this mass must contain either one carbon and two oxygens or three carbons.

$C + (2 \times O) = 12 + (2 \times 16) = 44$, so one H is needed to account for the mass. The fragment lost is CO_2H and the remaining fragment is C_7H_7.

$3 \times C = 36$ and so 9 Hs would be needed to give a mass of 45. It is not possible for such a fragment to occur in bonding terms.

77: formula C_6H_5

When ions are formed in the mass spectrometer, their bonds are weakened and they may break apart or fragment. Any charged fragments will reach the detector and appear in the mass spectrum. Notice that the two mass spectra are very different even though **A** and **B** have the same molecular formula.

CARE

It is a common mistake to confuse the peak of highest mass/charge ratio – the molecular ion or parent ion – with the tallest peak in the spectrum. This is called the **base peak**. The spectrum is usually scaled so that the relative abundance of the base peak is 100%. The molecular ion peak may have quite a low relative abundance as in the spectrum of **B**.

EXAM TIP

A fragment of mass 77 crops up quite often in mass spectra and it is worth remembering. It is C_6H_5 – a benzene ring less one hydrogen, called a phenyl group – and it often appears because of the stability of the benzene ring.

(c) What structural formulae would you predict for **A** and **B**?

One possibility for A is:

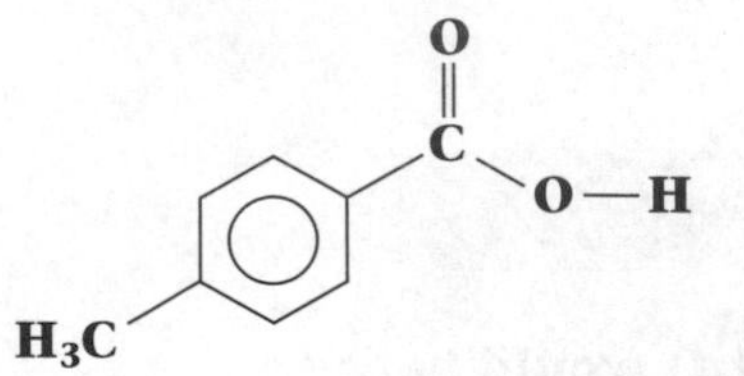

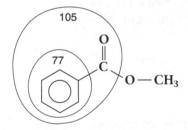

In fact the CH_3 group could be at any of the positions on the ring. This structure is supported by the peaks in the observed mass spectrum which can be accounted for by the breaking of just one bond.

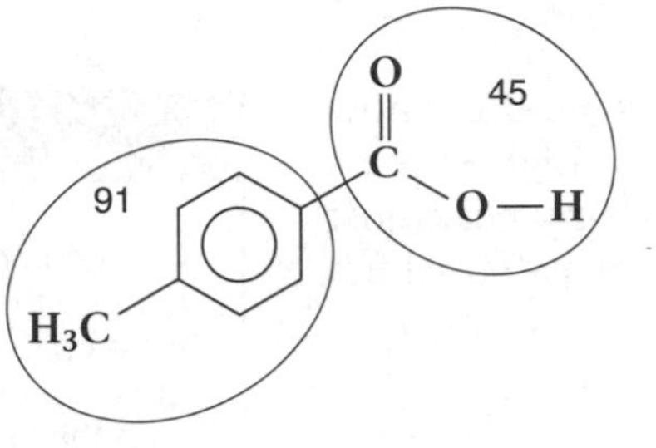

B is:

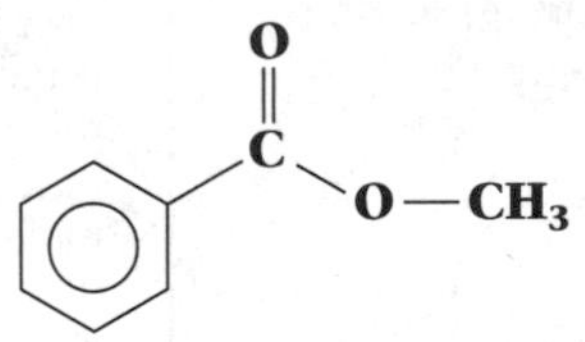

methyl benzoate

This is supported by looking at the main peaks in the mass spectrum.

The stem of the question gives a useful hint for this part of the question – see Exam Tip. So **B** is an ester containing a phenyl group. Esters have the functional group $—CO_2R$. The formula of a phenyl group is C_6H_5. These together add up to $C_7H_5O_2$ compared with the molecular formula of $C_8H_8O_2$. This means that R must be CH_3 and the structural formula is as shown above.

EXAM TIP

The phrase 'pleasant-smelling liquid' is often used by examiners to hint strongly at an ester. The smell of pear drops is the ester 3-methylbutyl ethanoate, for example.

Compound **A** is an isomer of **B** and so probably also contains a benzene ring. The most obvious isomer results from swapping the CH_3 group of the ester with one of the hydrogens on the ring. This would give a carboxylic acid. This would fit with the statement that **A** 'is moderately soluble in water'. The —CO_2H group can hydrogen bond to water but the benzene ring will reduce the solubility considerably. It is not possible to decide easily at which position on the ring the methyl group is situated.

(d)   Describe **two** tests or chemical reactions in which the behaviour of **A** and **B** would differ.

A will give an acid reaction to indicators; it will fizz when sodium carbonate solution is added and it will neutralise a significant amount of sodium hydroxide. B will do neither of these.

If you have got this far, this part should be easy as **A** is an acid and **B** is an ester.

A number of other answers are possible based on the differences between the reactions of acids and esters.

WHY DOES IT MATTER?

Mass spectrometers are among the chemist's most important tools – see Fig. 6.3a. They work by first converting the sample into positive ions. One way of doing this – called **electron impact** – is to bombard the sample with an electron beam which will knock out an electron from the sample molecule. Other methods include bombardment with positive ions – **chemical ionisation** – or with a beam of fast-moving inert gas atoms – **fast atom bombardment**. The ions are then accelerated by an electric field and deflected into a circular path by a magnetic field so that they hit a detector. Heavier ions need a stronger magnetic field to deflect them into the detector and so the mass of the sample molecule can be found from the size of the magnetic field.

Electron impact also tends to fragment the molecules of the sample and the masses of these fragments are detected too. The masses of fragments can help to unravel the structure of a compound as illustrated in the question above.

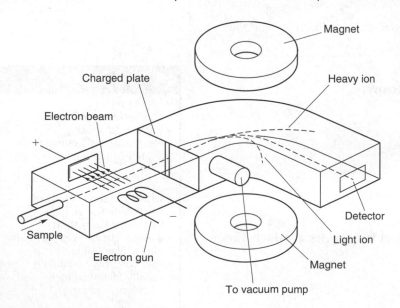

Fig. 6.3a   One type of mass spectrometer.

An unknown compound can be identified provided that the mass spectrum of the compound has been run before on the same instrument. A computer can be used to match its mass spectrum with a database, rather like matching fingerprints. This method is used to identify explosives from traces left on, for example, clothing, car seats or the hold of a blown-up aircraft. It is also used to screen for drugs in the urine samples given by athletes. The urine will contain many substances which are first separated by gas chromatography and then fed directly into the mass spectrometer. This method is called GCMS.

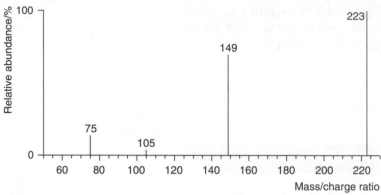

Fig. 6.3b   The mass spectrum of the explosive RDX, often used by terrorists.

6.4   (a)   Consider the following instrumental techniques used in structural analysis.

infra-red spectroscopy
ultra-violet/visible spectroscopy
proton magnetic resonance spectroscopy
mass spectrometry
X-ray diffraction

(i)   Which of these techniques uses electromagnetic radiation of the lowest frequency?

Proton magnetic resonance spectroscopy.

(ii)   Which technique uses electromagnetic radiation of the shortest wavelength?

X-ray diffraction.

Table 6.4a summarises the wavelengths and frequencies at which various instrumental techniques operate.

NOTE

The frequency (v) and wavelength (λ) of electromagnetic radiation are related by the equation

$c = v\lambda$

where c is the velocity of light. Radiation of high frequency has a short wavelength and vice versa. In ultra-violet/visible spectroscopy, radiation is usually identified by its wavelength in nm (10^{-9} m).

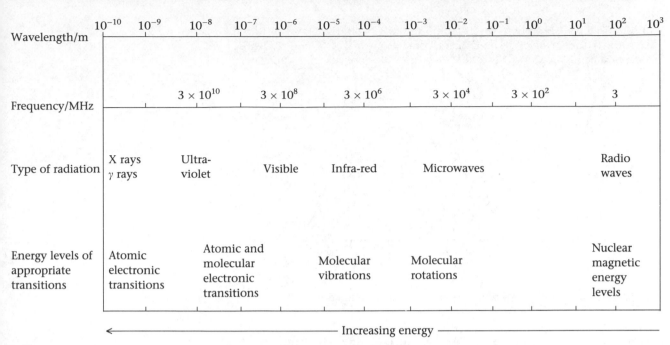

Wavelength/m	10^{-10}	10^{-9}	10^{-8}	10^{-7}	10^{-6}	10^{-5}	10^{-4}	10^{-3}	10^{-2}	10^{-1}	10^{0}	10^{1}	10^{2}	10^{3}
Frequency/MHz		3×10^{10}		3×10^{8}		3×10^{6}		3×10^{4}		3×10^{2}			3	
Type of radiation	X rays γ rays	Ultra-violet		Visible		Infra-red		Microwaves					Radio waves	
Energy levels of appropriate transitions	Atomic electronic transitions	Atomic and molecular electronic transitions			Molecular vibrations			Molecular rotations					Nuclear magnetic energy levels	

← ———————————————— Increasing energy ———————————————— →

Table 6.4a

(iii) In which technique is electromagnetic radiation absorbed by vibrating bonds?

Infra-red spectroscopy.

(iv) Which technique is the most generally appropriate for examining isotopically labelled molecules?

Mass spectrometry.

The mass of each atom is measured individually.

(v) Which technique can be used to determine the concentration of a compound in solution down to levels of approximately $10^{-5}\,\text{mol}\,\text{l}^{-1}$?

Ultra-violet/visible spectroscopy.

(vi) Indicate which technique often makes use of deuterated solvents and explain why this is so.

Proton nuclear magnetic resonance spectroscopy uses deuterated solvents – ones in which normal hydrogen atoms (^{1}H) are replaced by atoms of deuterium, the ^{2}H isotope of hydrogen. Deuterium atoms do not give rise to signals in the NMR spectrum and so do not swamp the spectrum of the sample.

(b) The visible absorption spectrum of the water effluent from a textile dyeworks is shown below. The effluent contains two dyes **X** and **Y**, whose absorption bands are shown.

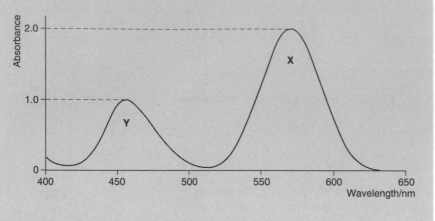

(i) Explain the process whereby a molecule absorbs energy from light.

Electromagnetic radiation of a particular frequency (or wavelength) may supply just the right amount of energy to promote an electron from one orbital to another.

Electrons in molecules (like those in atoms) exist in discrete energy levels (or orbitals). A beam of light shone into a sample will emerge missing those frequencies which correspond to the energy gaps between different orbitals.

(ii) Define the *absorbance* (A) of a solution in terms of the incident light intensity (I_0) and the emergent light intensity (I).

Absorbance (A) = $\log_{10} (I_0/I)$

This means that if light is absorbed, I is smaller than I_0, I_0/I becomes greater than 1 and $\log_{10} I_0/I$ is greater than 0. If, for example, the emergent light is ten times less intense than that going in, $I_0/I = 10$ and $\log_{10} (I_0/I = 1)$, giving an absorbance of 1.

(iii) State the expression which relates the molar extinction (absorption) coefficient (ε) of a substance to the absorbance of its solution at λ_{max}, the molar concentration (c) and the pathlength (l) of the solution.

$\epsilon = A/cl$

This means that a more concentrated solution of a sample will absorb more light (in fact the absorbance is proportional to the concentration) and so this method can be used to measure concentration.

(iv) Deduce from the above spectrum the *relative* concentrations of **X** and **Y** in the effluent, given that the ε_{max} value for **X** is three times that of **Y**.

For Y:

$A_Y = \epsilon_Y c_Y l$

$c_Y = A_Y/\epsilon_Y l$

Since the absorbance of **Y** = 1 from the spectrum,

$$c_Y = 1/\epsilon_Y l \qquad\qquad [1]$$

For X:

$$A_X = \epsilon_X c_X l$$

$$c_X = A_X/\epsilon_X l$$

Since the absorbance of **X** = 2,

$$c_X = 2/\epsilon_X l$$

But

$$\epsilon_X = 3\epsilon_Y$$

$$c_X = \tfrac{2}{3}\epsilon_Y l \qquad\qquad [2]$$

Comparing [1] and [2];

$$c_X = \tfrac{2}{3} c_Y$$

(v) On a particular occasion it was found that, in the absence of **X** and **Y**, the effluent was brown in colour. State why the absorbance measurements would give erroneous concentration values for **X** and **Y** when these two compounds are present in such an effluent. How might the error be minimised if a sample of the effluent containing no **X** or **Y** were available?

The brown colour would cause an absorption of light at many wavelengths which would be in addition to the absorption caused by X and Y.

So the absorption at 450 nm for example would be caused by both **Y** and the brown effluent.

One means of minimising the error would be to use a double-beam spectrophotometer.

Place brown effluent containing no X and Y in the 'blank' compartment. The instrument would subtract the absorbance of the brown effluent from that of the sample plus effluent, leaving the spectrum of the sample only.

WHY DOES IT MATTER?

Acid–base indicators

Indicators are substances which change colour depending on the pH of the solution in which they are dissolved. The colour change is brought about when one of the substances accepts a proton in acidic solution. For example phenolphthalein exists in an alkaline solution of pH 13 as [I], while in acid solution (pH 1) it accepts a proton and rearranges to exist as [II]. These two species have the UV/visible spectra shown in Figs. 6.4a and b.

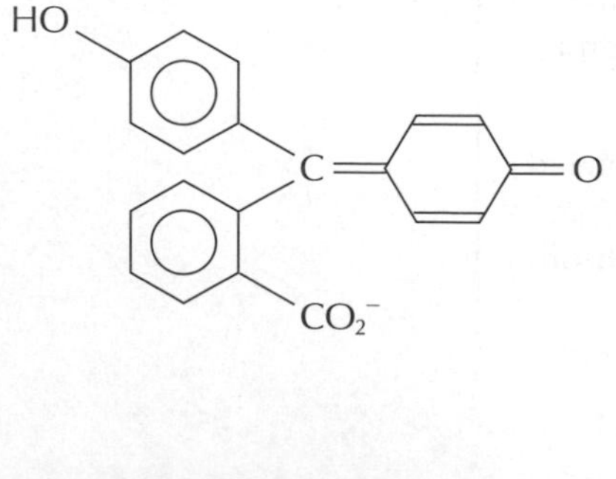

[I] Phenolphthalein
in alkaline solution

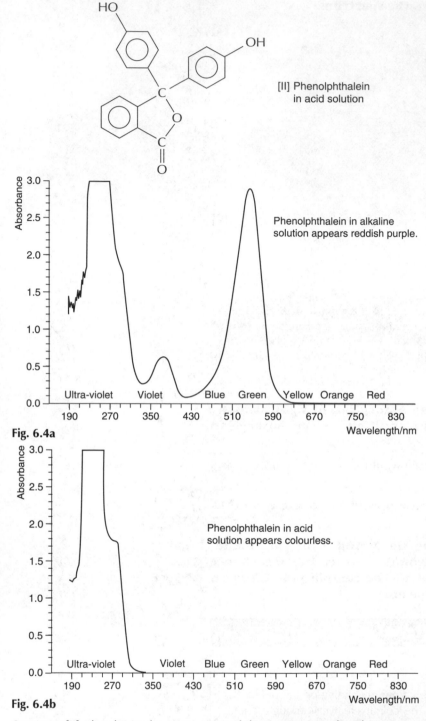

[II] Phenolphthalein
in acid solution

Fig. 6.4a

Phenolphthalein in alkaline
solution appears reddish purple.

Fig. 6.4b

Phenolphthalein in acid
solution appears colourless.

Structure [II] absorbs in the UV region of the spectrum only. This means that all the colours of visible light pass through it and it appears colourless. So phenolphthalein is colourless in acid solution. Structure [I] absorbs in the UV region, too. It also absorbs in the green/yellow and to a lesser extent in the blue/violet regions of the visible spectrum. This means that red and some blue light pass through and it appears reddish purple. So phenolphthalein is reddish purple in alkaline solution.

The reason that [I] absorbs in the visible region is that it has an extended delocalised system of alternating double and single bonds – a conjugated system – which [II] does not have. Conjugated systems often absorb visible light and cause species containing them to be coloured.

7 SYNOPTIC QUESTIONS

Synoptic questions are ones which cover a number of different areas of the syllabus.

What you should know now

❏ How to calculate oxidation numbers.

❏ How to calculate the number of moles in a specified quantity of:

an aqueous solution

a gas.

❏ How to draw Hess's law cycles to calculate ΔH for a reaction.

❏ Catalysts speed up reactions by reducing the activation energy.

❏ Isotopes are forms of an element with different numbers of neutrons in their atomic nuclei.

❏ The idea of entropy (S) as a measure of disorder in chemical systems.

❏ The relationship $\Delta S_{surroundings} = -\dfrac{\Delta H_{reaction}}{T}$.

❏ Weak acids are acids which dissociate partially in aqueous solution.

❏ The expression for the acid dissociation constant, K_a, of a weak acid.

❏ [ ] is used to represent the concentration of a species in $mol\,dm^{-3}$.

❏ $pH = -\log_{10}[H_3O^+(aq)]$ or the equivalent $pH = -\log_{10}[H^+(aq)]$.

❏ Shapes of molecules can be predicted by considering the number of groups of electrons in the outer shell of each atom.

Exam questions

7.1 This question is about sulphur dioxide. In the laboratory, the adsorption of sulphur dioxide may be demonstrated by passing sulphur dioxide through active charcoal, but this is not a practical method in industry. One way of tackling the acid rain problem, used in West German power stations, is to pass the waste gases containing sulphur dioxide through an aqueous suspension of limestone.
The overall reaction is

$$2SO_2(g) + 2CaCO_3(s) + O_2(g) \rightarrow$$
$$2CaSO_4(aq) + 2CO_2(aq)$$

Gypsum is then crystallised out as $CaSO_4.2H_2O(s)$. 1.2 million tonnes of gypsum are produced per year in West Germany by this method.

(a) Describe the changes you would expect to *see* when sulphur dioxide is passed through an aqueous suspension of limestone.   1 mark

(b) Sulphur dioxide can be detected by the reduction of $Cr_2O_7^{2-}$ ions to Cr^{3+} ions.
(i) Describe the changes you would expect to see when sulphur dioxide is passed through a solution of $Cr_2O_7^{2-}$ ions.
(ii) What is the oxidation number of chromium in the ion $Cr_2O_7^{2-}$?
(iii) In the reaction, the oxidation number of sulphur increases from +4 to +6. Suggest the likely product of the oxidation of sulphur dioxide and hence deduce the equation for this reaction.   5 marks

(c) (i) What does the term *adsorption* mean?
(ii) You are provided with a small cylinder containing sulphur dioxide. Draw a labelled diagram to show how you could measure out $100\,cm^3$ of sulphur dioxide and then find the proportion adsorbed by $10\,g$ of active charcoal.
(iii) How would you know when adsorption was complete?   5 marks

(d) The amount of sulphur dioxide adsorbed by the active charcoal can also be determined by a titration method.
The $10\,g$ of active charcoal containing the sulphur dioxide was added to $1000\,cm^3$ of iodine solution of concentration $0.00500\,mol$ of I_2 per dm^3.
$20.0\,cm^3$ portions of this solution were then titrated with $0.0100\,mol\,dm^{-3}$ sodium thiosulphate solution: $11.6\,cm^3$ were required for complete reaction.
The relevant equations are:

$$SO_2(g) + I_2(aq) + 2H_2O(l) \rightarrow$$
$$2I^-(aq) + SO_4^{2-}(aq) + 4H^+(aq)$$
$$I_2(aq) + 2S_2O_3^{2-}(aq) \rightarrow 2I^-(aq) + S_4O_6^{2-}(aq)$$

(i) Calculate the number of moles of sodium thiosulphate in $11.6\,cm^3$ of its solution.
(ii) Calculate the total number of moles of iodine, I_2, which reacted with the sodium thiosulphate.
(iii) Deduce the total number of moles of iodine which reacted with the sulphur dioxide.
(iv) Hence calculate the number of moles of sulphur dioxide present in $10\,g$ of active charcoal.   4 marks

(e) A power station produces 55 000 tonnes of gypsum per year: 1 tonne $= 10^3\,kg$.
(i) How many moles of gypsum are produced a year? (Relative atomic masses: H = 1, O = 16, S = 32, Ca = 40)
(ii) What volume of sulphur dioxide was absorbed in the production of 55 000 tonnes of gypsum? (1 mol of sulphur dioxide at this temperature has a volume of $24\,dm^3$.)   2 marks

(f) Suggest a use for the gypsum produced by this method.   1 mark

[L (Nuffield), '90]

7.2 Current estimates of world reserves of coal and oil suggest that we are likely to run short of oil long before coal deposits are exhausted. So chemists have turned their attention to ways of producing hydrocarbons from coal. One method which already exists (it was used in Germany during the Second World War and more recently in South Africa) is based on the Fischer–Tropsch process.

Carbon monoxide and hydrogen are the reactants in the Fischer–Tropsch process. A mixture of the two gases, the feedstock for the process, can be made by allowing coal to react with steam and air. Carbon monoxide and hydrogen do not react together on their own to produce hydrocarbons, and the Fischer–Tropsch process requires a catalyst. Quite complex mixtures of hydrocarbons can result, but the product consists mainly of hydrocarbons of between 5 and 8 carbon atoms per molecule if a cobalt catalyst, with traces of thorium oxide, is used.

(a) The equation for the production of octane from carbon monoxide and hydrogen is given in equation [1].

$$8CO(g) + 17H_2(g) \rightarrow C_8H_{18}(g) + 8H_2O(g) \qquad [1]$$

Use the following enthalpy changes of formation

$$\Delta H_f^{\ominus}: CO(g) = -111\,kJ\,mol^{-1},$$
$$C_8H_{18}(g) = -250\,kJ\,mol^{-1},$$
$$H_2O(g) = -242\,kJ\,mol^{-1}$$

to calculate the enthalpy change for equation [1].   3 marks

(b) (i) Explain why a large negative change in the entropy of the system (ΔS_{sys}) would be expected to accompany the reaction shown in equation [1].

2 marks

(ii) The total entropy change for this reaction must be positive, and therefore the change in the entropy of the surroundings (ΔS_{surr}) must be positive in this case. Explain why ΔS_{surr} is positive.

2 marks

(c) The reaction in equation [1] does not take place in the absence of a catalyst. Explain why a catalyst is needed.

2 marks

(d) The catalyst is made by depositing cobalt metal powder on granules of an inert material such as silicon dioxide. Describe one advantage of using a catalyst in this form.

2 marks

In order to have greater control over the products of the Fischer–Tropsch process, and to help design more effective catalysts, chemists have been carrying out research into the mechanisms of the reactions involved. The first step is thought to involve formation of carbon and hydrogen atoms which are bonded to the catalyst surface. The carbon and hydrogen atoms then react together to produce CH and CH_2 groups, which are also bonded to the catalyst, as shown in equation [2].

Chemists working at Sheffield Universidty have proposed that CH and CH_2 then react together to produce $CH{=}CH_2$ and that this group is the basis of the chain building process. Addition of a CH_2 group, followed by isomerisation, takes chain building one step further, and the chain continues to grow by successive additions of CH_2, followed by isomerisation as shown in equation [3]. (Again, all the groups are bonded to the catalyst.)

To check their proposal, the Sheffield chemists performed Fischer–Tropsch reactions to which they added small traces of $CH{=}CH_2$ radicals in which both carbons had been replaced with the ^{13}C isotope. Since almost all the hydrocarbon product contained two ^{13}C atoms, they assumed that their mechanism was correct.

(e) Explain the meaning of the term *isotope*.

1 mark

(f) Write down the numbers of protons, neutrons and electrons in an atom of the ^{13}C isotope.

1 mark

(g) Draw a full structural formula for a $CH{=}CH_2$ group which is bonded to the metal catalyst. (You should represent this as $M{-}CH{=}CH_2$.) Indicate on your drawing the approximate values you would expect for the bond angles in this structure.

2 marks

(h) Use this example to explain how the use of isotopes helps chemists to increase their understanding of chemical processes.

3 marks

[O&C (Salters'), '94]

7.3 Sulphur dioxide is present in the atmosphere from natural and industrial sources.

(a) When hydrogen sulphide in gases emitted from volcanoes mixes with air, it reacts with oxygen:

$$2H_2S(g) + 3O_2(g) \rightarrow 2H_2O(l) + 2SO_2(g)$$

(i) Use the following data to calculate the enthalpy change for this reaction (data at 298 K):

$$\Delta H_f^{\ominus} [H_2S(g)] = -20.6\,kJ\,mol^{-1}$$
$$\Delta H_f^{\ominus} [H_2O(l)] = -285.8\,kJ\,mol^{-1}$$
$$\Delta H_f^{\ominus} [SO_2(g)] = -296.8\,kJ\,mol^{-1}$$

(ii) Use your answer from (i) to calculate the entropy change in the surroundings at 298 K.

(iii) Calculate the entropy change in the system from the following data:

$$S^{\ominus}[H_2S(g)] = +205.7\,J\,mol^{-1}\,K^{-1}$$
$$S^{\ominus}[O_2(g)] = +205.0\,J\,mol^{-1}\,K^{-1}$$
$$S^{\ominus}[SO_2(g)] = +248.1\,J\,mol^{-1}\,K^{-1}$$
$$S^{\ominus}[H_2O(l)] = +69.9\,J\,mol^{-1}\,K^{-1}$$

(iv) Your answer to (iii), for the entropy change in the system, should include a positive or negative sign. Explain how this chemical change results in an entropy change which has this particular sign.

8 marks

(b) Sulphur dioxide dissolves in water droplets in the atmosphere and reacts to form an acidic solution of sulphurous acid.

$$H_2SO_3(aq) + H_2O(l) \rightleftharpoons H_3O^+(aq) + HSO_3^-(aq)$$

The equilibrium constant (dissociation constant) for this reaction, K_a, has a value of $1 \times 10^{-2}\,mol\,dm^{-3}$.

(i) Write down the expression for the equilibrium constant, K_a, for this reaction, including appropriate brackets and equilibrium subscripts.

(ii) A water droplet in the atmosphere above an industrial area was found to have a pH of 4.

If this acidic pH is due entirely to dissolved sulphur dioxide, use your answer to (i) to calculate the concentration, in $mol\,dm^{-3}$, of sulphurous acid in the water droplet.

$$pH = -lg\,[H_3O^+(aq)]$$

(iii) The dissolved sulphur dioxide is gradually oxidised by the air to sulphuric acid; K_a for sulphuric acid has a value of about $1 \times 10^2\,mol\,dm^{-3}$.

Will the pH of the water droplet increase, decrease or stay the same? Explain your answer.

8 marks

[L (Nuffield), specimen]

Answers

7.1 This question is about sulphur dioxide. In the laboratory, the adsorption of sulphur dioxide may be demonstrated by passing sulphur dioxide through active charcoal, but this is not a practical method in industry. One way of tackling the acid rain problem, used in West German power stations, is to pass the waste gases containing sulphur dioxide through an aqueous suspension of limestone.

The overall reaction is

$$2SO_2(g) + 2CaCO_3(s) + O_2(g) \rightarrow 2CaSO_4(aq) + 2CO_2(aq)$$

Gypsum is then crystallised out as $CaSO_4.2H_2O(s)$. 1.2 million tonnes of gypsum are produced per year in West Germany by this method.

(a) Describe the changes you would expect to *see* when sulphur dioxide is passed through an aqueous suspension of limestone.

The white cloudy suspension of calcium carbonate would go clear as it is replaced by an aqueous solution of calcium sulphate.

Note that the question asks about *changes* you would *see*, so make sure your answer describes the appearance of the liquid before and after. 'Calcium sulphate is formed' would not get full marks as it does not refer to changes nor does it say what you would see.

(b) Sulphur dioxide can be detected by the reduction of $Cr_2O_7^{2-}$ ions to Cr^{3+} ions.
(i) Describe the changes you would expect to see when sulphur dioxide is passed through a solution of $Cr_2O_7^{2-}$ ions.

An orange solution containing dichromate ions would turn green.

Again, the question asks about *changes* you would *see*.

(ii) What is the oxidation number of chromium in the ion $Cr_2O_7^{2-}$?

+VI

The + sign is important; do not leave it out.

Oxygen is normally —II in its compounds (except peroxide, superoxides and compounds with fluorine) so the seven oxygens contribute −14. The sum of the oxidation numbers must add up to the charge on the ion (−2), so the two chromiums together must contribute +12, i.e. +VI each. Oxidation numbers are normally written in Roman numerals to distinguish them from ionic charges.

(iii) In the reaction, the oxidation number of sulphur increases from +4 to +6. Suggest the likely product of the oxidation of sulphur dioxide and hence deduce the equation for this reaction.

The most likely species of sulphur(VI) is sulphate(VI) ions.

This is the unbalanced equation:

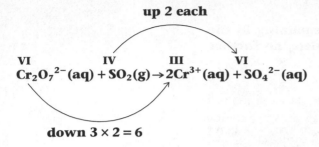

up 2 each

$$\overset{VI}{Cr_2O_7{}^{2-}}(aq) + \overset{IV}{SO_2}(g) \rightarrow 2\overset{III}{Cr^{3+}}(aq) + \overset{VI}{SO_4{}^{2-}}(aq)$$

down 3 × 2 = 6

$Cr_2O_7{}^{2-}$ **has two Cr atoms so there must be two Cr^{3+} ions on the right. Each Cr's oxidation number goes down by three giving a total drop of six which must be matched by an increase of six from the sulphurs. As each sulphur goes up by two, three sulphurs must be involved in the reaction.**

$$Cr_2O_7{}^{2-}(aq) + 3SO_2(g) \rightarrow 2Cr^{3+}(aq) + 3SO_4{}^{2-}(aq)$$

This is still unbalanced; there are 13 oxygens on the left and 12 on the right. This suggests that there must be a water molecule on the right and this requires two H^+ ions on the left. The equation is now balanced:

$$2H^+(aq) + Cr_2O_7{}^{2-}(aq) + 3SO_2(g) \rightarrow 2Cr^{3+}(aq) + 3SO_4{}^{2-}(aq) + H_2O(l)$$

Check that there are the same numbers of each type of atom on both sides and that the net charges on either side are the same. In this case the net charge on both sides is zero.

Note that the word 'deduce' indicates that you should show some working.

(a) (i) What does the term *adsorption* mean?

Adsorption is the formation of a layer of gas molecules on a solid surface.

The molecules are held either by chemical bonds (chemisorption) or intermolecular forces (physisorption).

(ii)   You are provided with a small cylinder containing sulphur dioxide. Draw a labelled diagram to show how you could measure out 100 cm³ of sulphur dioxide and then find the proportion adsorbed by 10 g of active charcoal.

A suitable apparatus is shown. 100 cm³ of gas is measured into one syringe (after flushing it out several times to displace any air). The gas is then passed to and fro from syringe to syringe over the charcoal to allow adsorption to take place.

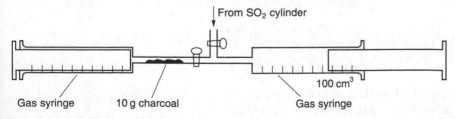

Fig. 7.1a

(iii) How would you know when adsorption was complete?

As adsorption takes place, the volume of gas remaining in the syringes would drop. When adsorption is complete, no further volume drop would occur.

(d) The amount of sulphur dioxide adsorbed by the active charcoal can also be determined by a titration method.

The 10 g of active charcoal containing the sulphur dioxide was added to 1000 cm³ of iodine solution of concentration 0.00500 mol of I_2 per dm³.

20.0 cm³ portions of this solution were then titrated with 0.0100 mol dm⁻³ sodium thiosulphate solution: 11.6 cm³ were required for complete reaction.

The relevant equations are:

$$SO_2(g) + I_2(aq) + 2H_2O(l) \rightarrow 2I^-(aq) + SO_4^{2-}(aq) + 4H^+(aq)$$

$$I_2(aq) + 2S_2O_3^{2-}(aq) \rightarrow 2I^-(aq) + S_4O_6^{2-}(aq)$$

(i) Calculate the number of moles of sodium thiosulphate in 11.6 cm³ of its solution.

The number of moles of solute concentration M mol dm⁻³ in V cm³ of solution is given by

$$\text{number of moles} = \frac{M \times V}{1000}$$

So in 11.6 cm³ of 0.0100 mol dm⁻³ sodium thiosulphate solution

$$\text{number of moles} = \frac{0.0100 \times 11.6}{1000}$$
$$= 1.16 \times 10^{-4} \text{ mol}$$

(ii) Calculate the total number of moles of iodine, I_2, which reacted with the sodium thiosulphate.

From the second equation, two moles of thiosulphate react with one mole of iodine, so the number of moles of iodine was 5.80×10^{-5} mol in a 20 cm³ portion of the solution.

So in the whole 1000 cm³ of solution, there were $50 \times 5.80 \times 10^{-5}$ mol = 2.90×10^{-3} mol of iodine.

(iii) Deduce the total number of moles of iodine which reacted with the sulphur dioxide.

In the 1000 cm³ of the original iodine solution there were 5.00×10^{-3} mol of iodine.

So $(5.00 - 2.90) \times 10^{-3} = 2.10 \times 10^{-3}$ mol of iodine was used up by reaction with sulphur dioxide.

(iv) Hence calculate the number of moles of sulphur dioxide present in 10 g of active charcoal.

The first equation tells us that 1 mol of iodine reacts with 1 mol of sulphur dioxide, so the charcoal must have adsorbed 2.10×10^{-3} mol of sulphur dioxide.

As a check on your answer, use the fact that 1 mol of any gas under room conditions has a volume of $24\,000\,cm^3$. 2.10×10^{-3} mol of sulphur dioxide would have a volume of $24\,000\,cm^3 \times 2.10 \times 10^{-3}\,cm^3 = 50.4\,cm^3$. This is a reasonable answer since we started with $100\,cm^3$ of gas.

> (e) A power station produces $55\,000$ tonnes of gypsum per year: 1 tonne $= 10^3\,kg$.
> (i)   How many moles of gypsum are produced per year?
> (Relative atomic masses: $H = 1$, $O = 16$, $S = 32$, $Ca = 40$)

The relative molecular mass of gypsum ($CaSO_4.2H_2O$) is $40 + 32 + (4 \times 16) + (2 \times 18) = 172$.

We have $55\,000 \times 1000 \times 1000 = 55 \times 10^9\,g$ of gypsum.

This is $\dfrac{55 \times 10^9}{172} = 3.19767 \times 10^8 = 3.2 \times 10^8$ mol.

> (ii)   What volume of sulphur dioxide was absorbed in the production of $55\,000$ tonnes of gypsum?
> (1 mol of sulphur dioxide at this temperature has a volume of $24\,dm^3$.)

The equation at the beginning of the question tells us that 1 mol of sulphur dioxide produces 1 mol of gypsum. So $3.2 \times 10^8 \times 24 = 7.7 \times 10^9\,dm^3$ of sulphur dioxide is absorbed.

> (f)   Suggest a use for the gypsum produced by this method.

Gypsum can be partially dehydrated to form plaster of Paris which is used for setting broken limbs and making plasterboard.

NOTE

The least number of significant figures in the data for this part of the question (e) is two, i.e. 24 in 24 000, so the answer is quoted to two significant figures. In part (d) the answer should be quoted to three significant figures, as this was the least accurate information.

WHY DOES IT MATTER?

Acid rain

Sulphur compounds are present in small amounts in fossil fuels. When the fuel is burnt in a power station the sulphur is converted to sulphur dioxide which is discharged into the atmosphere with the other waste or flue gases. In the atmosphere sulphur dioxide reacts with water and oxygen to become sulphuric acid, which later falls as rain.

$$SO_2(g) + \tfrac{1}{2}O_2(g) + H_2O(l) \rightarrow H_2SO_4(aq)$$

This acid rain is believed to be causing the death of fish in lakes and also of trees but it is difficult to be certain as there are so many variables to consider. There are a number of approaches to 'flue gas desulphurisation', one of which is described in the question. Another approach to the problem is to burn only coal which is low in sulphur. Coal, being a natural material, differs significantly in its composition depending on where it is found.

Sulphur dioxide is not the only contributor to acid rain. Another culprit is nitrogen oxides, NO_x, which are formed at high temperatures by the combination of nitrogen and oxygen from the air. These, too, react with oxygen and water vapour, to form nitric acid in this case.

7.2 Current estimates of world reserves of coal and oil suggest that we are likely to run short of oil long before coal deposits are exhausted. So chemists have turned their attention to ways of producing hydrocarbons from coal. One method which already exists (it was used in Germany during the Second World War and more recently in South Africa) is based on the Fischer–Tropsch process.

Carbon monoxide and hydrogen are the reactants in the Fischer–Tropsch process. A mixture of the two gases, the feedstock for the process, can be made by allowing coal to react with steam and air. Carbon monoxide and hydrogen do not react together on their own to produce hydrocarbons, and the Fischer–Tropsch process requires a catalyst. Quite complex mixtures of hydrocarbons can result, but the product consists mainly of hydrocarbons of between 5 and 8 carbon atoms per molecule if a cobalt catalyst, with traces of thorium oxide, is used.

(a) The equation for the production of octane from carbon monoxide and hydrogen is given in equation [1].

$$8CO(g) + 17H_2(g) \rightarrow C_8H_{18}(g) + 8H_2O(g) \qquad [1]$$

Use the following enthalpy changes of formation

$$\Delta H_f^{\ominus}: CO(g) = -111 \text{ kJ mol}^{-1},$$
$$C_8H_{18}(g) = -250 \text{ kJ mol}^{-1},$$
$$H_2O(g) = -242 \text{ kJ mol}^{-1}$$

to calculate the enthalpy change for equation [1].

This question may look daunting at first as you will probably not have covered the Fischer–Tropsch process in your course. However, all the question asks you to do is apply chemical principles to the process rather than know anything about it.

All values in kJ mol^{-1}

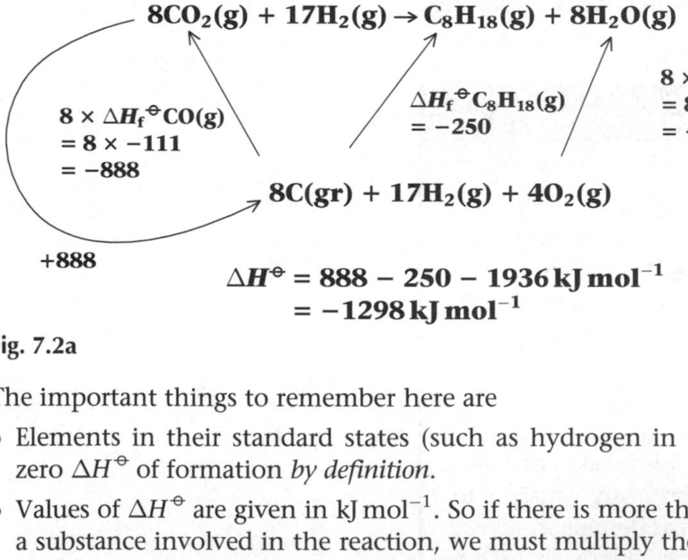

Fig. 7.2a

The important things to remember here are

- Elements in their standard states (such as hydrogen in this case) have zero $\Delta H^{\ominus}$ of formation *by definition*.

- Values of $\Delta H^{\ominus}$ are given in kJ mol^{-1}. So if there is more than one mole of a substance involved in the reaction, we must multiply the value of $\Delta H^{\ominus}$ by the appropriate number – eight for carbon monoxide in this example.

- When we reverse the direction of a chemical process, we reverse the sign of $\Delta H^{\ominus}$.

- To save time and space, write 'all values in kJ mol^{-1}' above the enthalpy cycle but remember to state the sign and units of the answer.

(b) (i) Explain why a large negative change in the entropy of the system (ΔS_{sys}) would be expected to accompany the reaction shown in equation [1].

Gases have high entropies as their molecules are arranged at random. This reaction has 25 mol of gas on the left and only 9 mol on the right so there is an increase in order of the system, i.e. a decrease in entropy, as we go from left to right.

The reaction is strongly exothermic, i.e. it transfers enthalpy (heat) from system to the surroundings. This enthalpy increase of the surroundings leads to an increase in disorder (entropy) of the surroundings.

In order for a reaction to be feasible the total entropy change of system plus surroundings must be positive.

Transferring heat to the surroundings means that the surroundings have more quanta of energy to be distributed between the appropriate energy levels. Thus there are more ways of arranging the system and it therefore has more entropy.

[$\Delta S = k\,\Delta \ln w$, where k is Boltzmann's constant (1.38×10^{-23} J K^{-1}) and w is the number of ways of arranging the system.]

Catalysts speed up reactions by making possible an alternative pathway with a lower activation energy. This reaction must have a high activation energy so that there is insufficient energy available to climb the activation energy 'hill' unless its height is reduced by using a catalyst.

The situation is shown in Fig. 7.2b.

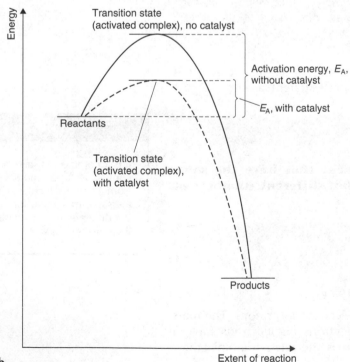

Fig. 7.2b

(d)   The catalyst is made by depositing cobalt metal powder on granules of an inert material such as silicon dioxide. Describe one advantage of using a catalyst in this form.

Powders have a large surface area to volume to ratio so the maximum amount of catalyst will be exposed to the reactants if it is used in this form.

In order to have greater control over the products of the Fischer–Tropsch process, and to help design more effective catalysts, chemists have been carrying out research into the mechanisms of the reactions involved. The first step is thought to involve formation of carbon and hydrogen atoms which are bonded to the catalyst surface. The carbon and hydrogen atoms then react together to produce CH and CH_2 groups, which are also bonded to the catalyst, as shown in equation [2].

$$ C \quad H \longrightarrow CH \xrightarrow[\text{H atom}]{\text{further}} CH_2 \qquad [2] $$

catalyst surface

Chemists working at Sheffield University have proposed that CH and CH_2 then react together to produce $CH{=}CH_2$ and that this group is the basis of the chain building process. Addition of a CH_2 group, followed by isomerisation, takes chain building one step further, and the chain continues to grow by successive additions of CH_2, followed by isomerisation as shown in equation [3]. (Again, all the groups are bonded to the catalyst.)

$$ CH_2 \quad CH{=}CH_2 \longrightarrow CH_2{-}CH{=}CH_2 \xrightarrow{\text{isomerisation}} CH{=}CH{-}CH_3 \xrightarrow{\text{etc.}} \qquad [3] $$

catalyst surface

To check their proposal, the Sheffield chemists performed Fischer–Tropsch reactions to which they added small traces of $CH{=}CH_2$ radicals in which both carbons had been replaced with the ^{13}C isotope. Since almost all the hydrocarbon product contained two ^{13}C atoms, they assumed that their mechanism was correct.

Do not be put off by the long and daunting section of text in the question at this point: the questions which follow are quite straightforward.

(e)   Explain the meaning of the term *isotope*.

Isotopes are forms of the same element that have the same atomic number, or proton number, but different numbers of neutrons.

(f)   Write down the number of protons, neutrons and electrons in an atom of the ^{13}C isotope.

Protons, 6; neutrons, $13 - 6 = 7$; electrons 6.

All neutral atoms have the same number of electrons as protons. The mass of the atom is due to the protons and neutrons (as electrons have a negligible mass) so the number of neutrons is the relative atomic mass minus the atomic number.

CARE

It is a common mistake to confuse isotopes and isomers. Isomers are molecules with the same molecular formula but different arrangements of their atoms in space.

(g)   Draw a full structural formula for a $CH=CH_2$ group which is bonded to the metal catalyst. (You should represent this as $M—CH=CH_2$.) Indicate on your drawing the approximate values you would expect for the bond angles in this structure.

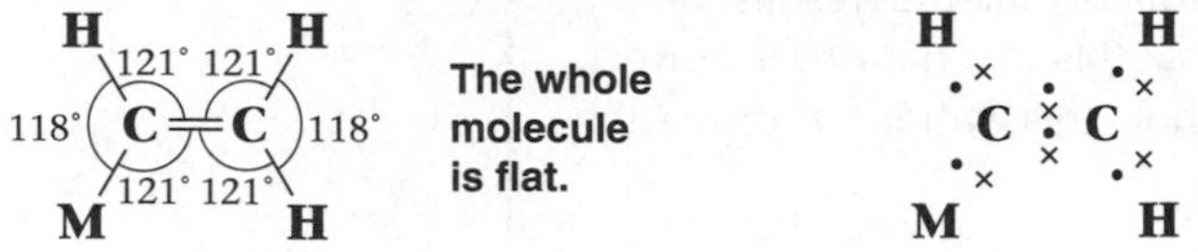

Fig. 7.2c

Each carbon has three groups of electrons around it (two single bonds and a double bond). These repel each other and the molecule adopts a flat trigonal shape to allow the groups of electrons to be as far apart as possible. The ideal bond angle would be 120° for all the bonds but the group of four electrons in the double bond repels more strongly than the pairs in each of the single bonds so the angle between the single bonds is 'squeezed' down by a couple of degrees. You would not be expected to know the exact angles but it is important that you realise that the H—C—H angle will be less than 120° and the H—C=C angle will be more than 120°.

(h)   Use this example to explain how the use of isotopes helps chemists to increase their understanding of chemical processes.

Isotopes can be useful in following what happens to particular atoms in chemical reactions. All isotopes of a particular element have the same numbers of protons and electrons. Therefore their electron arrangements are identical and they will react chemically in the same way. However, because they differ in mass, different isotopes can be detected separately using a mass spectrometer (or by measuring radioactivity if one of the isotopes is radioactive). The fact that almost all the product contained two ^{13}C atoms suggests that the $CH=CH_2$ radical remains intact during the reaction.

WHY DOES IT MATTER?

Isotopic labelling

A classic example of the use of isotopes to investigate reaction mechanisms is the acid-catalysed hydrolysis of esters.

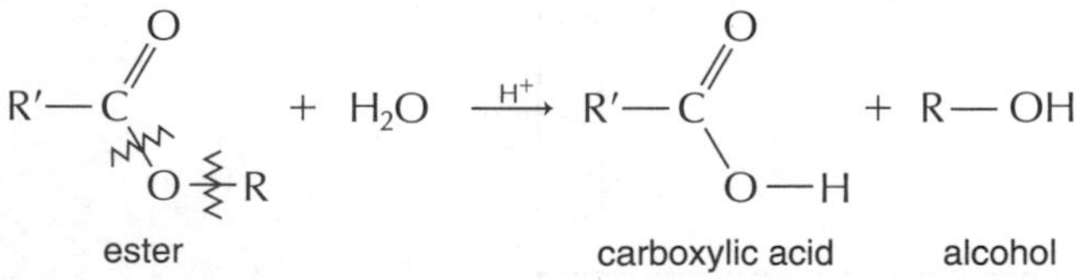

There are two possibilities: either the C—O bond or the O—R bond breaks.

If water containing the ^{18}O isotope is used in the reaction, the ^{18}O appears in the acid rather than in the alcohol. This suggests that the R—O bond stays intact and the C—O bond breaks. This is consistent with the mechanism in which the C=O of the ester is first protonated by H^+ ions from the acid catalyst and the resulting positive ions are attacked by water acting as a nucleophile. This is followed by a rearrangement and loss of ROH as a leaving group. See Fig. 7.2d. Notice that the results of the isotopic labelling experiment do not *prove* that this mechanism is correct. However, if the ^{18}O appeared in the alcohol, this would *disprove* the above mechanism.

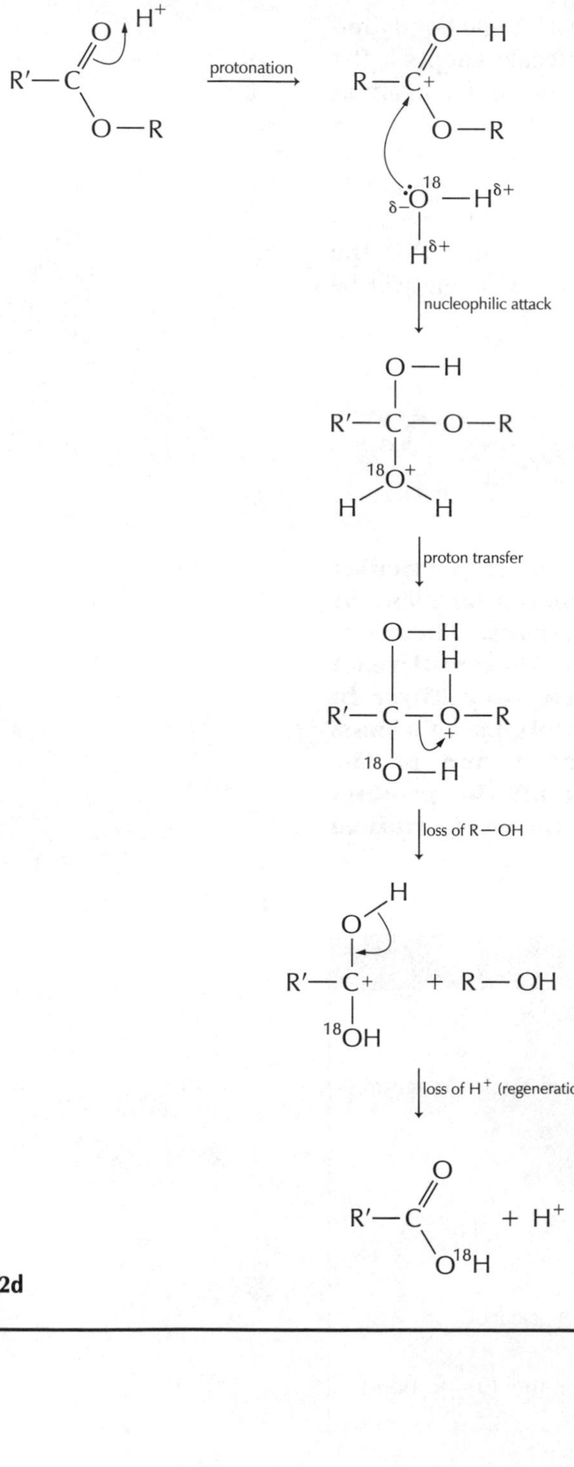

Fig. 7.2d

7.3 Sulphur dioxide is present in the atmosphere from natural and industrial sources.

(a) When hydrogen sulphide in gases emitted from volcanoes mixes with air, it reacts with oxygen:

$$2H_2S(g) + 3O_2(g) \rightarrow 2H_2O(l) + 2SO_2(g)$$

(i) Use the following data to calculate the enthalpy change for this reaction (data at 298 K).

$\Delta H_f^{\ominus} [H_2S(g)] = -20.6 \, kJ \, mol^{-1}$

$\Delta H_f^{\ominus} [H_2O(l)] = -285.8 \, kJ \, mol^{-1}$

$\Delta H_f^{\ominus} [SO_2(g)] = -296.8 \, kJ \, mol^{-1}$

All values in kJ mol^{-1}

$2H_2S(g) + 3O_2(g) \rightarrow 2H_2O(l) + 2SO_2(g)$

$2 \times \Delta H_f^{\ominus}(H_2S)$
$= 2 \times -20.6$
$= -41.2$

$2 \times \Delta H_f^{\ominus}(H_2O)$
$= 2 \times -285.8$
$= -571.6$

$2 \times \Delta H_f^{\ominus}(SO_2)$
$= 2 \times -296.8$
$= -593.6$

$+41.2$

$2H_2(g) + 2S(s) + 3O_2(g)$

$\Delta H_f^{\ominus} = 41.2 + -571.6 - 593.6 \, kJ \, mol^{-1}$
$\Delta H_f^{\ominus} = -1124.0 \, kJ \, mol^{-1}$

Remember to give the sign and the units.

The calculation is based on Hess's law which states that the enthalpy change of a chemical reaction is independent of the route by which it occurs. In this case $\Delta H^{\ominus}$ for the direct reaction is the same as that obtained by first converting the starting materials into the elements in their standard states and then converting the elements into the products.

CARE

Hess's law calculations lead to some common mistakes.

1. Forgetting to multiply the values of $\Delta H_f^{\ominus}$ for H_2S, SO_2 and H_2O by 2. All ΔHs are given per mole, so if two molecules are involved, the values must be multiplied by two.

2. Note that O_2 is an element in its standard state so its $\Delta H_f^{\ominus}$ is zero. This is not likely to confuse you in this question where the values of $\Delta H_f^{\ominus}$ you need are given, but it often puzzles students in questions where they have to look up data.

3. Remember to change the sign when you reverse the direction of an arrow (as for the value of $\Delta H_f^{\ominus}$ for H_2S in this case).

(ii) Use your answer from (i) to calculate the entropy change in the surroundings at 298 K.

Using the relationship

$$\Delta S_{surroundings} = \frac{-\Delta H_{reaction}}{T}$$

$$\Delta S_{surroundings} = \frac{(1124.0 \times 1000)}{298}$$

$$= +3772 \, J \, mol^{-1} \, K^{-1}$$

Remember that T is in K, not °C. The factor of 1000 appears because enthalpy is measured in $kJ \, mol^{-1}$ and entropy in $J \, mol^{-1} \, K^{-1}$. Remember that the upper case K in $J \, mol^{-1} \, K^{-1}$ stands for kelvin while the lower case k in $kJ \, mol^{-1}$ stands for kilo.

The minus sign is there in the relationship because if the reaction gives out heat (is exothermic, ΔH negative) from the system to the surroundings then the entropy of the surroundings increases.

This is an exothermic reaction and heat energy (enthalpy) is transferred *from* the system (the vessel where the reaction takes place) *to* the surroundings.

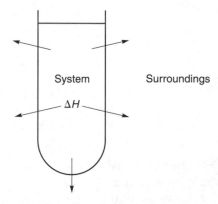

System   Surroundings

ΔH

An exothermic reaction transfers enthalpy from the system to the surroundings.

Fig. 7.3a

This means that after the reaction there are more quanta of energy in the surroundings than before and therefore more ways of arranging them, i.e. the surroundings are more disordered. The entropy of the surroundings has increased.

The more enthalpy that is transferred to the surroundings, the greater the increase in entropy. The extra quanta have proportionately more effect at low temperature where there are fewer quanta than at higher temperatures.

(iii) Calculate the entropy change in the system from the following data:

$S^{\ominus}[H_2S(g)] = +205.7 \, J \, mol^{-1} \, K^{-1}$

$S^{\ominus}[O_2(g)] = +205.0 \, J \, mol^{-1} \, K^{-1}$

$S^{\ominus}[SO_2(g)] = +248.1 \, J \, mol^{-1} \, K^{-1}$

$S^{\ominus}[H_2O(l)] = +69.9 \, J \, mol^{-1} \, K^{-1}$

$$\Delta S^{\ominus}_{syst} = \text{total } S^{\ominus}_{products} - \text{total } S^{\ominus}_{reactants}$$

$$= \{[2 \times S^{\ominus}(H_2O(l))] + 2 \times [S^{\ominus}(SO_2(g))]\}$$

$$- \{2 \times [S^{\ominus}(H_2S(g))] + 3 \times [S^{\ominus}(O_2(g))]\}$$

$$= \{(2 \times 69.9) + (2 \times 248.1)\} - \{(2 \times 205.7) + (3 \times 205.0)\}$$

$$= \{139.8 + 496.2\} - \{411.4 + 615.0\}$$

$$= 636.0 - 1026.4$$

$$\Delta S^{\ominus}_{syst} = -390.4 \, J \, mol^{-1} \, K^{-1}$$

NOTE

There is a possible ambiguity in the question here. The figure for $S^{\ominus}$ for oxygen could be given per mole of oxygen molecules or per mole of oxygen atoms. The figure is in fact that for a mole of molecules.

(iv)   Your answer to (iii), for the entropy change in the system, should include a positive or negative sign. Explain how this chemical change results in an entropy change which has this particular sign.

ΔS_{syst} **is negative. Entropy is a measure of randomness. The products (2 mol of gas and 2 mol of liquid) are less random (more ordered) than reactants (5 mol of gases).**

(b)   Sulphur dioxide dissolves in water droplets in the atmosphere and reacts to form an acidic solution of sulphurous acid.

$$H_2SO_3(aq) + H_2O(l) \rightleftharpoons H_3O^+(aq) + HSO_3^-(aq)$$

The equilibrium constant (dissociation constant) for this reaction, K_a, has a value of $1 \times 10^{-2} \, mol \, dm^{-3}$.
(i)   Write down the expression for the equilibrium constant, K_a, for this reaction, including appropriate brackets and equilibrium subscripts.

$$K_a = \frac{[H_3O^+(aq)]_{eqm} \times [HSO_3^-(aq)]_{eqm}}{[H_2SO_3(aq)]_{eqm}}$$

Note that there is no need to include $[H_2O(l)]$ in the expression as, being a pure liquid, its concentration cannot change and, by convention, it is incorporated into K_a.

CARE

Don't forget the state symbols and eqm subscripts. You are reminded to include these in this question but you may not always be.

Remember [ ] indicates the concentration of the species in the square brackets in $mol \, dm^{-3}$.

(ii)   A water droplet in the atmosphere above an industrial area was found to have a pH of 4.
If this acidic pH is due entirely to dissolved sulphur dioxide, use your answer to (i) to calculate the concentration, in $mol \, dm^{-3}$, of sulphurous acid in the water droplet.

$$pH = -lg[H_3O^+(aq)]$$

$$pH = - \log_{10} [H_3O^+(aq)]$$

$$4 = - \log_{10} [H_3O^+(aq)]$$

$$- 4 = \log_{10} [H_3O^+(aq)]$$

$$[H_3O^+(aq)] = 1 \times 10^{-4} \, mol \, dm^{-3}$$

Do this step using INV lg buttons on your calculator if you are at all unsure.

$$K_a = \frac{[H_3O^+(aq)]_{eqm} \times [HSO_3^-(aq)]_{eqm}}{[H_2SO_3(aq)]_{eqm}}$$

Since the equation for the dissociation tells us that formation of each H_3O^+ also involves formation of an HSO_3^- and there is no other source of either,

$$[H_3O^+(aq)]_{eqm} = [HSO_3^-(aq)]_{eqm}$$

So

$$1 \times 10^{-2} = \frac{1 \times 10^{-4} \times 1 \times 10^{-4}}{[H_2SO_3(aq)]_{eqm}}$$

$$[H_2SO_3(aq)]_{eqm} = 1 \times 10^{-8}/1 \times 10^{-2}$$

$$[H_2SO_3(aq)]_{eqm} = 1 \times 10^{-6} \, mol \, dm^{-3}$$

The equation for dissociation could equally well be written

$$H_2SO_3(aq) \rightleftharpoons H^+(aq) + HSO_3^-(aq)$$

and the equilibrium constant expression would then become

$$K_a = \frac{[H^+(aq)]_{eqm} \times [HSO_3^-(aq)]_{eqm}}{[H_2SO_3(aq)]_{eqm}}$$

This makes no difference to the answer.

(iii) The dissolved sulphur dioxide is gradually oxidised by the air to sulphuric acid; K_a for sulphuric acid has a value of about $1 \times 10^2 \, mol \, dm^{-3}$.
Will the pH of the water droplet increase, decrease or stay the same? Explain your answer.

It will decrease as K_a is larger for H_2SO_4 than for H_2SO_3, so there will be a greater $[H_3O^+(aq)]_{eqm}$. Since $pH = - lg[H_3O^+(aq)]$, the pH will decrease.

CARE

More acidic means *lower* pH; a common error is to assume the opposite.

WHY DOES IT MATTER?

Acid rain

A great deal has been written and said about acid rain in recent years suggesting that it is a recent phenomenon caused by industrial pollution. In fact rain is naturally acidic for a number of reasons.

1. Carbon dioxide in the atmosphere reacts with water to form carbonic acid, a weak acid.

2. There are naturally occurring sulphur compounds in the atmosphere as described in the question.

3. Nitrogen oxides are formed in the air by lightning strikes, where the high temperature allows nitrogen and oxygen to combine. These oxides react with oxygen and water to produce nitric acid.

The activities of mankind over the last century or so have undoubtedly increased the acidity of rain, however. Fig. 7.3b shows the acidity of a Scottish lake since the year 1300. This has been measured by examining the skeletons of organisms called diatoms found in the loch's sediment. Different types of diatoms thrive at different acidities.

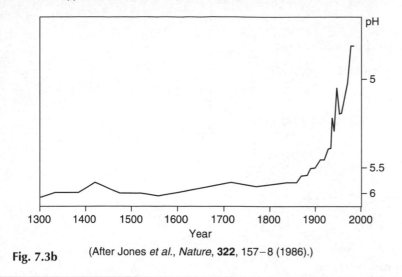

Fig. 7.3b   (After Jones *et al.*, *Nature*, **322**, 157–8 (1986).)

8 COMPREHENSION QUESTIONS

What you should know now

❏ Benzene (C_6H_6) is unusually stable and unreactive owing to its circular system of delocalised π electrons, called an aromatic system. See Fig. 8.1a.

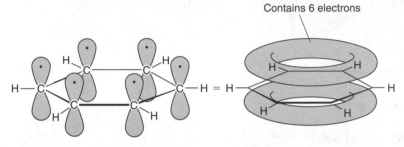

Contains 6 electrons

The cyclic delocalised π system in benzene

Fig. 8.1a

❏ Although they are unsaturated, benzene and its substituted derivatives tend to undergo electrophilic substitution reactions, rather than addition reactions which would destroy the π system.

❏ S_N1 reactions involve formation of intermediate positive ions (carbocations).

❏ Cations are positive ions, anions are negative.

❏ How to write structural formulae for organic compounds given the systematic names.

❏ The difference between thermoplastic and thermosetting polymers.

❏ How to draw 'dot–cross' diagrams for simple ionic and covalent compounds.

❏ How to predict the shapes of simple molecules using electron pair repulsion theory.

❏ The reactions of alcohols and of carbonyl compounds (aldehydes and ketones).

❏ How to write simple balanced equations with state symbols.

❏ How to do calculations of reacting quantities.

Exam questions

8.1 Benzene, C_6H_6, is an aromatic hydrocarbon. The six carbon atoms form a regular hexagon, and the six hydrogen atoms attached to these carbon atoms are also in the plane of this ring. Although benzene can be written as cyclohexa-1,3,5-triene (**A**) it rarely shows the reactions of alkenes. The usual type of reaction is the substitution of hydrogen by other groups. Thus, nitration takes place when benzene is treated with a mixture of nitric and sulphuric acids; nitrobenzene is formed by an electrophilic attack upon the ring. The benzene system of π electrons is cyclic and is apparently unusually stable. Although the addition of hydrogen to benzene will occur, giving firstly cyclohexadiene, then cyclohexene and lastly cyclohexane, it is the first step which is the most difficult. Once the π system in benzene has been attacked the product (cyclohexadiene) behaves as an alkene and undergoes rapid and exothermic addition. Similarly, bromine adds to benzene in the presence of light to give the adduct $C_6H_6Br_6$; once the benzene π system has been disrupted, further addition takes place quickly.

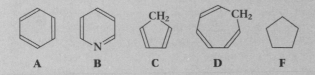

A   B   C   D   F

The stability of the benzene aromatic system is associated with the continuous, flat, cyclic system of six π electrons. Other systems exist with this structural feature. In the pyridine molecule (**B**) one of the CH systems has been replaced by a nitrogen atom, but the flat six-membered ring is still present and the six π electrons are spread over it. Cyclopenta-1,3-diene (**C**) is not aromatic, although it is flat, because the π system is not cyclic and because there are only four π electrons associated with it. Cyclohepta-1,3,5-triene (**D**) is also not aromatic. Although there are six π electrons associated with it, they are not in a continuous system because of the $-CH_2-$ system which disrupts it. However, if one of the hydrogen atoms of this $-CH_2-$ group is removed with its bonding electrons, the resulting carbocation (**E**) is aromatic. Salts containing **E** dissolve in water and **E** shows unusual stability, reacting only slowly with the solvent.

(a) (i) Suggest one substitution reaction of pyridine, and give the structure of a possible product.
(ii) Pyridine also reacts with HCl, although benzene does not. Explain this observation, and give the structure of the product of this reaction.   10 marks

(b) (i) Write an equation for the formation of the cation (**E**) from cyclohepta-1,3,5-triene (**D**).

(ii) Explain why this cation could be described as aromatic, and by reference to the S_N1 mechanism of hydrolysis of alkyl halides show how the behaviour of **E** suggests that it is unusually stable.   10 marks

(c) Cyclopentane (**F**) may be dried by distillation over sodium metal; water reacts readily with the metal, but the cyclopentane does not. Cyclopenta-1,3-diene (**C**) reacts readily with sodium to give hydrogen and the cyclopentadienyl anion, $C_5H_5^-$.
(i) Write an equation for the reaction of cyclopenta-1,3-diene with sodium.
(ii) Explain why the cyclopentadienyl anion may be regarded as aromatic.
(iii) Suggest whether the cyclopentadienyl *cation* is aromatic, and justify your answer.   10 marks

[L, '93]

8.2 Read the following passage about Biopol straight through, and then more carefully, in order to answer the following questions.

(a) Write a summary in continuous prose, in no more than 150 words, to describe the current manufacture, properties and uses of Biopol.

Credit will be given for answers written in good English, using complete sentences and with the correct use of technical words. Numbers count as one word, as do standard abbreviations and hyphenated words. If you include chemical equations or formulae they do not count in the word total, nor does the title to your account. At the end of your account, state clearly the number of words you have used. There are penalties for the use of words in excess of 150.

(b) (i) Draw structural formulae for 3-hydroxybutanoic acid and 3-hydroxypentanoic acid. Show how one molecule of each could join together to form part of the polyester Biopol.
(ii) Predict whether the polymer Biopol would be a thermosetting or a thermoplastic polymer. Explain your answer.

Biopol: a New Biodegradable Plastic

Since the early 1950s, polymeric materials have revolutionised the way most people live in industrialised countries. Because of their versatility and relatively low cost, polymers are essential components of most consumer articles, with the total world production now in excess of 100 million tonnes per annum.

The development of new products has always been high on the list of priorities for the chemical industry. There are two main reasons for this: firstly to increase the UK share of the highly competitive world plastics market;

and secondly to combat the environmental problems associated with the growth of this market. Tough, durable, throw-away plastic products are difficult to dispose of – do you, for instance, bury them in a landfill site, or do you burn them in an incinerator? With the level of environmental concern at an all-time high, the general public is calling for materials with the toughness and flexibility of plastics, yet with the environmental 'friendliness' of non-plastics – that is, materials that are biodegradable.

Recently, ICI announced the launch of a new biodegradable plastic – 'Biopol'. This new plastic has all the durability, stability and water resistance of conventional plastics, but on disposal it is quickly and efficiently broken down into carbon dioxide and water.

'Biopol' polymers are linear polyesters of 3-hydroxybutanoic acid and 3-hydroxypentanoic acid. These acids are produced by the fermentation of a mixed feedstock of carbohydrate and organic acid by the naturally occurring bacterium *Alcaligenes eutrophus* which is widely distributed in soil, and fresh and salt waters, and is involved directly in the spoilage of food.

In the first stage of production, *Alcaligenes eutrophus* is inoculated into a growth medium containing glucose and mineral salts, at the specified temperature for optimum growth. During a period of rapid growth, biosynthesis of the polymer begins. Although it is strictly not a lipid, the polymer will be a component of the so-called lipid granules of the microorganism. At the end of the growth cycle the organisms have accumulated up to 80% of their dry mass as the polymer. The cells are then harvested and treated so that they break open. The crude polymer is removed by solvent extraction and then purified. What is in fact an energy reserve for the microorganism, like fat in mammals, now becomes a commercial plastic.

How does the microorganism produce the necessary monomers to form the polymer? As well as 2-hydroxypropanoic acid (lactic acid) being formed as a product of the fermentation of carbohydrates, other products are possible including 3-hydroxybutanoic acid. Chemists, by adding pentanoic acid to the growth medium, have produced Biopol with a range of compositions from 0–20% of 3-hydroxypentanoic acid.

Biopol has a variety of uses depending on its composition. Since it is a biocompatible compound, it can be used in a range of medical implants – e.g. by choosing the correct composition of the polymer, it is possible to produce slow release capsules for veterinary use, with the polymeric outer material slowly breaking down to release controlled amounts of drugs at the site of the implant.

Biopol products can be made using conventional polymer-processing technology, although care must be taken at all times because, as a polyester, Biopol is liable to be unstable at high temperatures. At the typical processing temperatures used for Biopol (185–190 °C) the

time the polymer is in the melt needs to be minimised. At temperatures above 205 °C the polymer degrades rapidly.

Although we will probably be using Biopol in a variety of ways in the future, most available literature concentrates on its biodegradability. So why does Biopol degrade so easily? For an answer to this we must remember that the substance is a polyester and contains compounds that are present in a range of living systems. On disposal to a suitable site and with the correct conditions, fungi and bacteria break down the Biopol in a few weeks. The first step is probably the hydrolysis of the ester linkages. After hydrolysis, the compounds produced are metabolised through a fatty acid oxidation cycle producing energy, carbon dioxide and water.

Will Biopol ever be cheap enough to replace non-degradable plastics such as poly(ethene)? The answer to this must be no. However, with effective 'green' advertising, consumers are probably prepared to pay a premium for this material. Also the gene responsible for synthesising poly-3-hydroxybutanoic acid was identified in 1987 and since then scientists have been attempting to introduce it into other organisms. It is possible that crop plants with this gene may soon be able to produce the polymer on a large scale, so reducing its cost.

Is the widespread adoption of biodegradable plastics the long-term answer for a waste-making industrialised society? What about the recyling of plastics? The answers to these questions lie in the future, but the production of Biopol is a step in the right direction.

(801 words)

(Adapted from an article by Dr. P.S. Phillips, Education in Chemistry, Volume 28, No. 5 September, 1991.)

[L (Nuffield), '93]

8.3 *Read the following passage and then answer the questions which follow. You are not expected to write extended answers to these questions.*

Boron, like carbon, forms a wide range of hydrides. The simplest of these, diborane, B_2H_6, is a gas at room temperature which inflames spontaneously in air. The shape of the molecule is similar to that of dimeric aluminium chloride and it is the simplest *electron-deficient* molecule. Its high molar enthalpy of combustion prompted interest in it as a rocket fuel but, although its properties were much investigated with this in mind, it was never used.

Diborane also shows some reactions similar to those of aluminium chloride in that donor ligands (such as ammonia and amines) cause the bridging B—H—B bonds to break, forming addition compounds. In Al_2Cl_6 only symmetrical bond breaking occurs forming addition compounds such as $AlCl_3N(CH_3)_3$, while ammonia and diborane react in a 2:1 mole ratio to give an addition compound of formula $B_2N_2H_{12}$ formed by unsymmetrical cleavage of the bridging bonds. This addition compound forms conducting solutions in inert solvents.

Of much greater importance are the borohydrides. The borohydride ion, BH_4^-, is isoelectronic with methane.

The borohydrides of the Group 1 metals are ionic solids which can be prepared by the reaction of the corresponding metal hydride with diborane in a non-aqueous solvent. Sodium borohydride is a versatile and selective reducing agent for organic molecules which brings about reduction by nucleophilic hydride ion transfer. It reduces ketones, acid chlorides and aldehydes but leaves alkene double bonds unaffected.

Sodium borohydride became commercially available in the early 1960s as a result of a process developed by Bayer in which the solids sodium tetraborate, $Na_2B_4O_7$, and silicon dioxide are reacted with sodium and hydrogen. The resulting mixture is extracted under pressure with liquid ammonia which is evaporated giving a 98% pure product as a powder. Borohydrides are extensively used in chemical plating processes. The conventional electroplating process is not universally applicable to all materials and chemical plating processes must be used instead. Sodium borohydride was introduced on an industrial scale in the early 1960s mainly for the deposition of corrosion-resistant, bright nickel films. Chemical plating also achieves a uniform thickness of deposit independent of the geometrical shape, however complicated. Nickel ions are reduced at pH 14 to give a nickel deposit which contains some boron and corresponds to the formula $Ni_{10}B_3$. The equation for the overall reaction is:

$$10NiCl_2 + 8NaBH_4 + 17NaOH + 3H_2O \rightarrow Ni_{10}B_3$$
$$+ 5NaB(OH)_4 + 20NaCl + 17.5H_2$$

(a) (i) Draw the 'dot and cross' representation of the electronic structure of the species BH_3.
(ii) Draw the structure of diborane. In the light of the bonding in this structure, suggest what is implied by the term 'electron-deficient' molecule'.

(iii) Suggest, and justify, a structure for the addition compound formed in the reaction between ammonia and diborane.   6 marks

(b) (i) Draw the shape of the borohydride ion and describe the nature of the bonding present.
(ii) What is meant by the term 'isoelectronic'?
(iii) Give the formula of a polyatomic cation which is isoelectronic with the borohydride ion.
(iv) What structural feature is common to isoelectronic species?   5 marks

(c) (i) Write an equation for the preparation of sodium borohydride.
(ii) Suggest why this reaction cannot be carried out in water and name a suitable inert solvent which could be used.
(iii) Suggest a mechanism for the reduction of ketones by sodium borohydride using butanone as an example.
(iv) What property would the product of the reaction in (iii) show which is not shown by butanone? Explain your answer.
(v) Suggest why sodium borohydride leaves alkene double bonds unaffected.   9 marks

(d) (i) Write an equation, with state symbols, for the combustion of diborane in oxygen.
(ii) Suggest **two** reasons why diborane was not used as a rocket fuel.   3 marks

(e) (i) Suggest **two** steps which might be taken in the Bayer process to increase the rate of the reaction.
(ii) Why must extraction of the product be carried out under pressure?   3 marks

(f) (i) Give **two** advantages of chemical plating over electroplating.
(ii) Calculate the mass of $Ni_{10}B_3$ which can be deposited by 1.0 kg of sodium borohydride.   4 marks

[L, '95]

Answers

8.1 Benzene, C_6H_6, is an aromatic hydrocarbon. The six carbon atoms form a regular hexagon, and the six hydrogen atoms attached to these carbon atoms are also in the plane of this ring. Although benzene can be written as cyclohexa-1,3,5-triene (**A**) it rarely shows the reactions of alkenes. The usual type of reaction is the substitution of hydrogen by other groups. Thus, nitration takes place when benzene is treated with a mixture of nitric and sulphuric acids; nitrobenzene is formed by an electrophilic attack upon the ring. The benzene system of π electrons is cyclic and is apparently unusually stable. Although the addition of hydrogen to benzene will occur, giving firstly cyclohexadiene, then cyclohexene and lastly cyclohexane, it is the first step which is the most difficult. Once the π system in benzene has been attacked the product (cyclohexadiene) behaves as an alkene and undergoes rapid and exothermic addition. Similarly, bromine adds to benzene in the presence of light to give the adduct $C_6H_6Br_6$; once the benzene π system has been disrupted, further addition takes place quickly.

The stability of the benzene aromatic system is associated with the continuous, flat, cyclic system of six π electrons. Other systems exist with this structural feature. In the pyridine molecule (**B**) one of the CH systems has been replaced by a nitrogen atom, but the flat six-membered ring is still present and the six π electrons are spread over it. Cyclopenta-1,3-diene (**C**) is not aromatic, although it is flat, because the π system is not cyclic and because there are only four π electrons associated with it. Cyclohepta-1,3,5-triene (**D**) is also not aromatic. Although there are six π electrons associated with it, they are not in a continuous system because of the —CH_2— system which disrupts it. However, if one of the hydrogen atoms of this —CH_2— group is removed with its bonding electrons, the resulting carbocation (**E**) is aromatic. Salts containing **E** dissolve in water and **E** shows unusual stability, reacting only slowly with the solvent.

(a) (i) Suggest one substitution reaction of pyridine, and give the structure of a possible product.

Any electrophilic substitution reaction of benzene that you know might be expected to occur with pyridine. For example:

1. **Reaction with the NO_2^+ ion, generated in a mixture of concentrated nitric and sulphuric acids to give nitropyridine.**

2. **Reaction with liquid bromine with an appropriate catalyst such as iron filings to give bromopyridine.**

3. **Reaction with concentrated sulphuric acid or a solution of sulphur trioxide in sulphuric acid ('fuming sulphuric acid') to give pyridine sulphonic acid.**

Pyridine is aromatic and would therefore be expected to react in the same sort of way as benzene with electrophiles, so the above are sensible suggestions.

In fact pyridine is somewhat less reactive than benzene as the electronegative nitrogen atom withdraws electrons from the π system, making it less easily attacked by electrophiles. So more vigorous conditions have to be used for substitution reactions of pyridine than of benzene. Substitution generally occurs at the 3-position although you would not be expected to know either of these facts for this question.

(ii)   Pyridine also reacts with HCl, although benzene does not. Explain this observation, and give the structure of the product of this reaction.

The nitrogen atom in pyridine has a lone pair making pyridine a base. It will therefore react with hydrochloric acid to form a salt in which the lone pair has accepted a proton.

Pyridine's basicity is another reason for it being less reactive to electrophiles than is benzene. With acidic electrophiles, pyridine tends to accept a proton and the resulting positive ion is less easily attacked by electrophiles which are themselves positive.

(b) (i) Write an equation for the formation of the cation (**E**) from cyclohepta-1,3,5-triene (**D**).

The passage states that the hydrogen must be removed with both its bonding electrons, i.e. as an H^- ion, not as an H atom nor as the more familiar H^+ ion.

(ii)   Explain why this cation could be described as aromatic, and by reference to the S_N1 mechanism of hydrolysis of alkyl halides show how the behaviour of **E** suggests that it is unusually stable.

Cycloheptatriene has six π electrons and hence could be aromatic. But in cycloheptatriene there is a CH_2 group which stops the π system being cyclic. If H^- is removed, leaving the cycloheptatrienyl cation, the resulting CH group will be hybridised sp^2 leaving an empty p-orbital which can participate in the π system making it cyclic – see Fig. 8.1b.

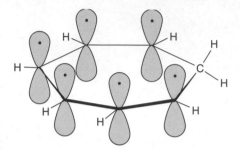

In cycloheptatriene the π system cannot be cyclic because of the CH_2 group which has no p-orbitals available.

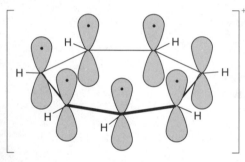

The cycloheptatrienyl cation has 6 π electrons in p-orbitals which can overlap to give a cyclic π system.

Fig. 8.1b

In the S_N1 hydrolysis of alkyl halides, the first (slow) step is the loss of a halide ion leaving a carbocation. This is rapidly attacked by water (a weak nucleophile), followed by loss of a proton to give an alcohol.

$$R\!-\!X \longrightarrow R^+ + X^- \xrightarrow{H_2O} R\!-\!\overset{H}{\underset{H}{\overset{+}{O}}} \longrightarrow ROH + H^+$$

The passage states that the cycloheptatrienyl cation reacts only slowly with the solvent (water) indicating that it is much more stable than alkyl carbocations.

(c)   Cyclopentane (**F**) may be dried by distillation over sodium metal; water reacts readily with the metal, but the cyclopentane does not. Cyclopenta-1,3-diene (**C**) reacts readily with sodium to give hydrogen and the cyclopentadienyl anion, $C_5H_5^-$.

(i)   Write an equation for the reaction of cyclopenta-1,3-diene with sodium.

$$\underset{C}{\overset{H \quad H}{\diagup \diagdown}} + Na \longrightarrow Na^+ + \underset{C^-}{\overset{H}{|}} + \tfrac{1}{2}H_2$$

This is analogous with the reaction of sodium with water

$$H_2O + Na \rightarrow Na^+ + OH^- + \frac{1}{2}H_2$$

(ii)   Explain why the cyclopentadienyl anion may be regarded as aromatic.

Cyclopentadiene has only four electrons in its π system (two from each double bond. Loss of H⁺ leaves behind both of the electrons in the C—H bond. The remaining C atom is hybridised sp² and the electron pair exists in a p-orbital which can contribute to a cyclic π system which now has six electrons.

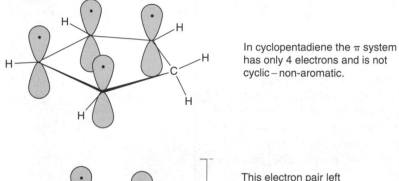

In cyclopentadiene the π system has only 4 electrons and is not cyclic – non-aromatic.

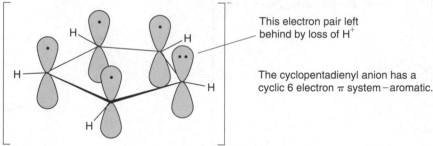

This electron pair left behind by loss of H⁺

The cyclopentadienyl anion has a cyclic 6 electron π system – aromatic.

Fig. 8.1c

(iii) Suggest whether the cyclopentadienyl *cation* is aromatic, and justify your answer.

The cyclopentadienyl cation would be formed from cyclopentadiene by loss of :H⁻, i.e. the H atom would be lost with the electron pair from the C—H bond. This leaves the cyclopentadienyl cation with only four electrons and therefore it cannot be aromatic.

WHY DOES IT MATTER?

Carcinogens

Six is not the only number of electrons which can give a compound the special stability which we describe as aromatic. The general rule is that $4n + 2$ (where n is a whole number) electrons are required in a flat, cyclic system of overlapping p-orbitals. Benzene and the other compounds in this question have six π electrons (i.e. $n = 1$). Naphthalene has 10 π ($n = 2$) electrons and anthracene and phenanthrene have 14 ($n = 3$) – see Fig. 8.1d.

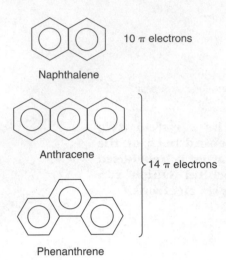

Fig. 8.1d

These are called polynuclear aromatic hydrocarbons. Many of the more complex ones are powerful carcinogens (cancer-causing chemicals). A particularly potent one is benzopyrene (Fig. 8.1e) which is found in a variety of places such as cigarette smoke, soot, car exhaust fumes, and even barbecued meat. These compounds are believed to cause cancers by forming compounds which can bond to DNA in cells and affect the DNA's ability to replicate itself accurately.

Fig. 8.1e

Benzopyrene

BY THE WAY

Benzene is often drawn  rather than to

emphasise that all the carbon–carbon bond lengths are the same and that there are no localised double bonds. Polynuclear aromatic hydrocarbons are sometimes drawn in the same way as in Fig. 8.1e. However the situation is not as simple as in benzene, as not all the carbon–carbon bond lengths are the same. The bond lengths in naphthalene, for example, are

0.142 nm
0.136 nm
0.142 nm
0.142 nm

Various types of notation have been used including

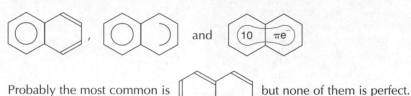

Probably the most common is ⬡⬡ but none of them is perfect.

8.2 Read the following passage about Biopol straight through, and then more carefully, in order to answer the following questions.

(a) Write a summary in continuous prose, in no more than 150 words, to describe the current manufacture, properties and uses of Biopol.

Credit will be given for answers written in good English, using complete sentences and with the correct use of technical words. Numbers count as one word, as do standard abbreviations and hyphenated words. If you include chemical equations or formulae they do not count in the word total, nor does the title to your account. At the end of your account, state clearly the number of words you have used. There are penalties for the use of words in excess of 150.

Biopol: a New Biodegradable Plastic

Since the early 1950s, polymeric materials have revolutionised the way most people live in industrialised countries. Because of their versatility and relatively low cost, polymers are essential components of most consumer articles, with the total world production now in excess of 100 million tonnes per annum.

The development of new products has always been high on the list of priorities for the chemical industry. There are two main reasons for this: firstly to increase the UK share of the highly competitive world plastics market; and secondly to combat the environmental problems associated with the growth of this market. Tough, durable, throw-away plastic products are difficult to dispose of – do you, for instance, bury them in a landfill site, or do you burn them in an incinerator? With the level of environmental concern at an all-time high, the general public is calling for materials with the toughness and flexibility of plastics, yet with the environmental 'friendliness' of non-plastics – that is, materials that are biodegradable.

Recently, ICI announced the launch of a new biodegradable plastic – 'Biopol'. This new plastic has all the durability, stability and water resistance of conventional plastics, but on disposal it is quickly and efficiently broken down into carbon dioxide and water.

'Biopol' polymers are linear polyesters of 3-hydroxybutanoic acid and 3-hydroxypentanoic acid. These acids are produced by the fermentation of a mixed feedstock of carbohydrate and organic acid by the naturally occurring bacterium *Alcaligenes eutrophus* which is widely distributed in soil, and fresh and salt waters, and is involved directly in the spoilage of food.

In the first stage of production, *Alcaligenes eutrophus* is inoculated into a growth medium containing glucose and mineral salts, at the specified temperature for optimum growth. During a period of rapid growth, biosynthesis of the polymer begins. Although it is strictly not a lipid, the polymer will be a component of the so-called lipid granules of the microorganism. At the end of the growth cycle the organisms have accumulated up to 80% of their dry mass as the polymer. The cells are then harvested and treated so that they break open. The crude polymer is removed by solvent extraction and then purified. What is in fact an energy reserve for the microorganism, like fat in mammals, now becomes a commercial plastic.

How does the microorganism produce the necessary monomers to form the polymer? As well as 2-hydroxypropanoic acid (lactic acid) being formed as a product of the fermentation of carbohydrates, other products are possible including 3-hydroxybutanoic acid. Chemists, by adding pentanoic acid to the growth medium, have produced Biopol with a range of compositions from 0–20% of 3-hydroxypentanoic acid.

Biopol has a variety of uses depending on its composition. Since it is a biocompatible compound, it can be used in a range of medical implants – e.g. by choosing the correct composition of the polymer, it is possible to produce slow release capsules for veterinary use, with the polymeric outer material slowly breaking down to release controlled amounts of drugs at the site of the implant.

Biopol products can be made using conventional polymer-processing technology, although care must be taken at all times because, as a polyester, Biopol is liable to be unstable at high temperatures. At the typical processing temperatures used for Biopol (185–190 °C) the time the polymer is in the melt needs to be minimised. At temperatures above 205 °C the polymer degrades rapidly.

Although we will probably be using Biopol in a variety of ways in the future, most available literature concentrates on its biodegradability. So why does Biopol degrade so easily? For an answer to this we must remember that the substance is a polyester and contains compounds that are present in a range of living systems. On disposal to a suitable site and with the correct conditions, fungi and bacteria break down the Biopol in a few weeks. The first step is probably the hydrolysis of the ester linkages. After hydrolysis, the compounds produced are metabolised through a fatty acid oxidation cycle producing energy, carbon dioxide and water.

Will Biopol ever be cheap enough to replace non-degradable plastics such as poly(ethene)? The answer to this must be no. However, with effective 'green' advertising, consumers are probably prepared to pay a premium for this material. Also the gene responsible for synthesising poly-3-hydroxybutanoic acid was identified in 1987 and since then scientists have been attempting to introduce it into other organisms. It is possible that crop plants with this gene may soon be able to produce the polymer on a large scale, so reducing its cost.

Is the widespread adoption of biodegradable plastics the long-term answer for a waste-making industrialised society? What about the recyling of plastics? The answers to these questions lie in the future, but the production of Biopol is a step in the right direction.

(801 words)

(Adapted from an article by Dr. P.S. Phillips, Education in Chemistry, Volume 28, No. 5 September, 1991)

Some exam boards set comprehension questions in which writing a summary forms a major part. Often, as in this example, you will be given a brief for the summary. Here you are asked to 'describe the current manufacture, properties and uses of Biopol'. It is important to stick to this brief: you will get no credit for describing the environmental issues regarding Biopol, for example.

You are also given a word limit for the summary. It is vital to stick to this as marks are deducted for exceeding the word count – typically at a rate of one mark for every five excess words. Summaries are usually marked by the examiners identifying a number of key points which they look for in the answer – usually about a dozen – and awarding one mark for each covered in the answer up to a maximum of, say, 10. There will also be a few marks for clear English and good style, so do not try to save words by missing out words such as 'the' and 'and'.

The best way to tackle these questions is to read the passage once in full and then again, looking for and marking key points (using a pencil so that you can change your mind) bearing in mind the brief. Having identified the key points, divide the number of key points into the word limit to give you an approximate idea of how many words to write on each point.

Key points for this summary are:

- **Biopol is made by fermenting a mixture of carbohydrates and organic acids.**

- **The bacterium *Alcaligenes eutrophus* is used.**

- **After the growth cycle, the cells are harvested and broken open.**

- **Crude polymer is separated by solvent extraction and purified.**

- **Processing takes place at between 185 °C and 190 °C.**

- **Biopol degrades rapidly above 205 °C.**

- **Biopol is durable, stable and water-resistant.**

- **Biopol is broken down by fungi and bacteria.**

- **The breakdown products are carbon dioxide and water.**

- **The first step of breakdown is the hydrolysis of ester linkages.**

- **An important use is for slow release capsules for drugs and for medical implants.**

Up to nine marks were available – one for each point.

> (b) (i) Draw structural formulae for 3-hydroxybutanoic acid and 3-hydroxypentanoic acid. Show how one molecule of each could join together to form part of the polyester Biopol.

3-Hydroxybutanoic acid is

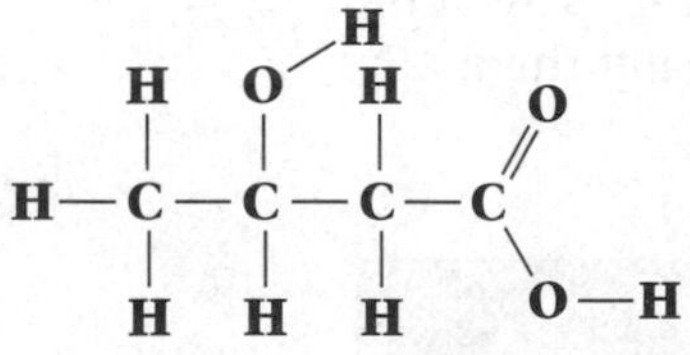

3-Hydroxypentanoic acid is

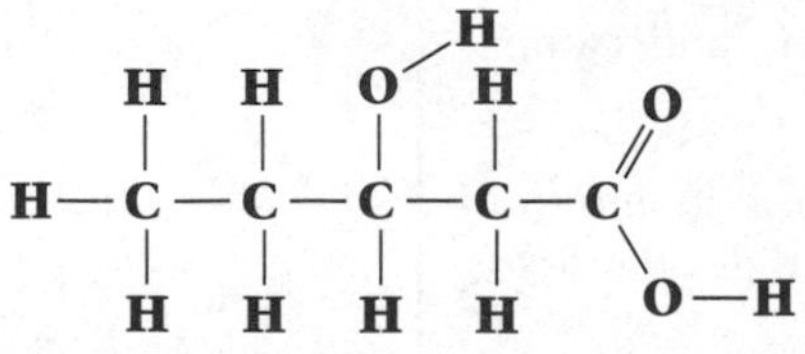

An ester can be formed by the reaction of the $-CO_2H$ group on one acid with the $-OH$ group on another with elimination of a molecule of water – see Fig. 8.2a.

EXAM TIP

Write your summary double-spaced. This will give you room to make corrections legibly without having to write the whole summary out again.

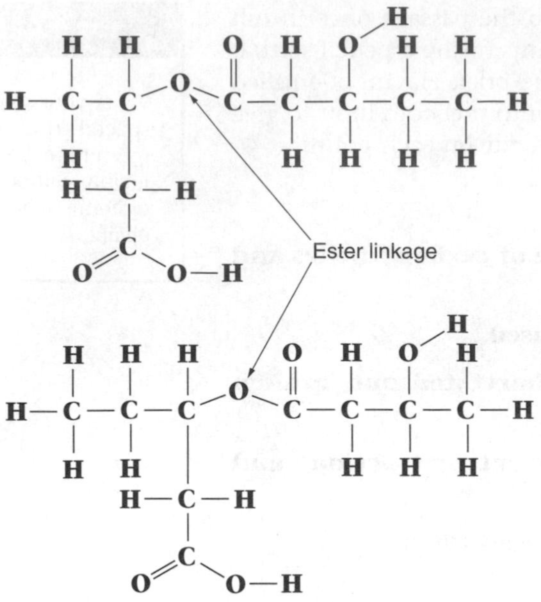

Both the dimers have a free —O—H group
and a free —CO₂H group for further reaction.

Fig. 8.2a

This can be done in two ways; the acid group of 3-hydroxybutanoic acid reacting with the alcohol group of 3-hydroxypentanoic acid or the acid group of 3-hydroxypentanoic acid reacting with the alcohol group of 3-hydroxybutanoic acid.

(ii) Predict whether the polymer Biopol would be a thermosetting or a thermoplastic polymer. Explain your answer.

Biopol is a thermoplastic as there are no covalent bonds forming cross-links between adjacent chains.

Thermoplastics soften on heating and harden again on cooling. This is because chains are held together by weak van der Waals bonds. Thermosetting polymers have covalent bonds which form cross-links between chains. On heating, they do not soften but eventually thermally decompose or char.

WHY DOES IT MATTER?

Disposing of plastics

Worldwide production of plastics is over 100 million tonnes annually, much of it for 'throwaway' applications such as packaging and carrier bags. There are a number of options for disposal:

- burying along with other rubbish in landfill sites – expensive in terms of land, and harmful products of decomposition may find their way into watercourses
- incineration – this can provide energy for schemes such as district heating
- pyrolysis (heating in the absence of air) – this can produce chemicals similar to crude oil which can be used as fuel or chemical feedstock

• recycling by melting down thermoplastics and remoulding them – this requires sorting of the plastics into different types first and it can only be done a limited number of times as heating the polymer chains degrades them

Biodegradable plastics are a possible solution to part of this problem. One solution already being used is to incorporate a biodegradable material such as starch into a conventional plastic. In the environment the starch will degrade rapidly leaving fragments of the conventional plastic. This approach is used in some 'biodegradable' carrier bags.

8.3   *Read the following passage and then answer the questions which follow. You are not expected to write extended answers to these questions.*

Boron, like carbon, forms a wide range of hydrides. The simplest of these, diborane, B_2H_6, is a gas at room temperature which inflames spontaneously in air. The shape of the molecule is similar to that of dimeric aluminium chloride and it is the simplest *electron-deficient* molecule. Its high molar enthalpy of combustion prompted interest in it as a rocket fuel but, although its properties were much investigated with this in mind, it was never used.

Diborane also shows some reactions similar to those of aluminium chloride in that donor ligands (such as ammonia and amines) cause the bridging B—H—B bonds to break, forming addition compounds. In Al_2Cl_6 only symmetrical bond breaking occurs forming addition compounds such as $AlCl_3N(CH_3)_3$, while ammonia and diborane react in a 2:1 mole ratio to give an addition compound of formula $B_2N_2H_{12}$ formed by unsymmetrical cleavage of the bridging bonds. This addition compound forms conducting solutions in inert solvents.

Of much greater importance are the borohydrides. The borohydride ion, BH_4^-, is isoelectronic with methane.

The borohydrides of the Group 1 metals are ionic solids which can be prepared by the reaction of the corresponding metal hydride with diborane in a non-aqueous solvent. Sodium borohydride is a versatile and selective reducing agent for organic molecules which brings about reduction by nucleophilic hydride ion transfer. It reduces ketones, acid chlorides and aldehydes but leaves alkene double bonds unaffected.

Sodium borohydride became commercially available in the early 1960s as a result of a process developed by Bayer in which the solids sodium tetraborate, $Na_2B_4O_7$, and silicon dioxide are reacted with sodium and hydrogen. The resulting mixture is extracted under pressure with liquid ammonia which is evaporated giving a 98% pure product as a powder. Borohydrides are extensively used in chemical plating processes. The conventional electroplating process is not universally applicable to all materials and chemical plating processes must be used instead. Sodium borohydride was introduced on an industrial scale in the early 1960s mainly for the deposition of corrosion-resistant, bright nickel films. Chemical plating also achieves a uniform thickness of deposit independent of the geometrical shape, however complicated. Nickel ions are reduced at pH 14 to give a nickel deposit which contains some boron and corresponds to the formula $Ni_{10}B_3$. The equation for the overall reaction is:

$$10NiCl_2 + 8NaBH_4 + 17NaOH + 3H_2O \rightarrow Ni_{10}B_3$$

$$+ 5NaB(OH)_4 + 20NaCl + 17.5H_2$$

(a) (i)
Draw the 'dot and cross' representation of the electronic structure of the species BH_3.

H
·×
B ×H
·×
H

Note that BH_3 has only six electrons in its outer shell: it is electron-deficient.

(ii) Draw the structure of diborane. In the light of the bonding in this structure, suggest what is implied by the term '*electron-deficient* molecule'.

H H H
  B B
H H H

In fact the bridging hydrogens are above and below the plane of the rest of the atoms making the shape around each of the boron atoms approximately tetrahedral – see Fig. 8.3a.

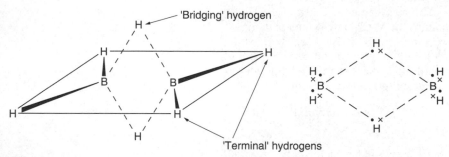

Fig. 8.3a

Each of the bridging hydrogens forms two bonds (one with each boron) yet it has only two electrons.

(iii) Suggest, and justify, a structure for the addition compound formed in the reaction between ammonia and diborane.

BH_4^- $BH_2(NH_3)_2^+$

'Unsymmetrical cleavage of the bridging bonds' suggests that both bridging hydrogens go to one of the boron atoms, leaving a BH_4^- ion and a BH_2^+ ion. BH_4^- is stable but BH_2^+ would be severely electron-deficient and would react with two ammonia molecules which would form dative covalent bonds to boron using their lone pairs. The result would be an ionic compound $BH_2(NH_3)_2^+BH_4^-$. The fact that the addition compound forms conducting solutions is a clue that it must be ionic.

Note that the word 'suggest' implies that you are not expected to know the 'right' answer but that a sensible application of chemistry knowledge and principles will gain credit.

(b) (i) Draw the shape of the borohydride ion and describe the nature of the bonding present.

$$\left[\begin{array}{c} H \\ | \\ \overset{\displaystyle H}{\underset{\displaystyle H}{\overset{|}{B}}} \cdots H \end{array} \right]^{-}$$

The ion will be a regular tetrahedron.

There are four covalent bonds one of which is formally a dative covalent bond between a BH_3 molecule and the lone pair of an H^- ion. But as each bond contains a pair of electrons, this makes no difference to the shape.

(ii) What is meant by the term 'isoelectronic'?

Having the same number of electrons as.

'Iso' means 'equal' as in 'isomer' which means 'a compound having the same formula'.

(iii) Give the formula of a polyatomic cation which is isoelectronic with the borohydride ion.

NH_4^+. The N and the B atoms in these compounds both have eight electrons in their outer shells (10 in all).

'Polyatomic' means 'having more than one atom in the molecule'. Cations are positive ions: they are attracted to the cathode.

(iv) What structural feature is common to isoelectronic species?

They have the same shape as they have the same number of electron pairs in their outer shells.

(c) (i) Write an equation for the preparation of sodium borohydride.

$2NaH(s) + B_2H_6(g) \rightarrow 2NaBH_4(s)$

(ii) Suggest why this reaction cannot be carried out in water and name a suitable inert solvent which could be used.

Sodium hydride reacts with water (to form hydrogen and sodium hydroxide). A fairly polar solvent would be needed to dissolve $NaBH_4$ which is ionic, possibly ether (ethoxyethane).

(iii) Suggest a mechanism for the reduction of ketones by sodium borohydride using butanone as an example.

$$\underset{\underset{\displaystyle C_2H_5}{\overset{\displaystyle H_3C}{\diagdown}}}{\overset{\delta+}{C}}=O^{\delta-} \longrightarrow CH_3-\underset{\underset{\displaystyle C_2H_5}{\overset{\displaystyle |}{}}}{\overset{\overset{\displaystyle H}{\overset{\displaystyle |}{}}}{C}}-O^- \xrightarrow{H^+} CH_3-\underset{\underset{\displaystyle C_2H_5}{\overset{\displaystyle |}{}}}{\overset{\overset{\displaystyle H}{\overset{\displaystyle |}{}}}{C}}-O-H$$

The passage says that the reaction proceeds via the hydride ion ($:H^-$).

The first step will be a nucleophilic attack by the H^- ion on the electron-deficient $C^{\delta+}$ of the C=O. The resulting $-O^-$ species will pick up a proton from the solvent to produce an alcohol.

(iv) What property would the product of the reaction in (iii) show which is not shown by butanone? Explain your answer.

The product is an alcohol. So any reaction of an alcohol which is not shown by a ketone would do. For example, reaction with sodium to produce hydrogen, reaction with orange acidified dichromate ions to produce green Cr^{3+} ions, reaction with phosphorus pentachloride to give fumes of hydrogen chloride.

(v) Suggest why sodium borohydride leaves alkene double bonds unaffected.

The attacking species is $:H^-$ which is a nucleophile and attacks electron deficient $C^{\delta+}$. Alkenes have an electron-rich double bond and are attacked by electrophiles.

(d) (i) Write an equation, with state symbols, for the combustion of diborane in oxygen.

$B_2H_6(g) + 3O_2(g) \rightarrow B_2O_3(s) + 3H_2O(l)$

The formula of boron oxide is B_2O_3 by analogy with that of aluminium oxide.

(ii) Suggest **two** reasons why diborane was not used as a rocket fuel.

The passage states that diborane inflames spontaneously in air which would make it difficult and dangerous to handle. It is a gas at room temperature which means that it would be bulky to store or that it would have to be cooled for storage as a liquid.

(e) (i) Suggest **two** steps which might be taken in the Bayer process to increase the rate of the reaction.

Any two methods of increasing the rate would do:

- **increasing the temperature**
- **increasing the pressure of hydrogen**
- **grinding the solids more finely to increase their surface area.**

Using a catalyst is not as good a suggestion because it is not specific enough. You would really need to name a suitable catalyst – which you are unlikely to know!

(ii)   Why must extraction of the product be carried out under pressure?

Ammonia is a very volatile liquid (it has a low boiling point) and would boil away if not kept under pressure.

In fact ammonia boils at 240 K ($-33\,°C$) at atmospheric pressure.

(f) (i)   Give **two** advantages of chemical plating over electroplating.

The passage states that chemical plating is applicable to a wider range of materials than electroplating and that it produces a uniform thickness of plating even on complex shapes.

You would not be expected to know this, merely select the relevant information from the passage.

(ii)   Calculate the mass of $Ni_{10}B_3$ which can be deposited by 1.0 kg of sodium borohydride.

From the equation, 8 mol of sodium borohydride produces 1 mol of $Ni_{10}B_3$.

The relative molecular masses are

$NaBH_4$: $23 + 11 + (4 \times 1) = 38$

$Ni_{10}B_3$: $10 \times 59 + (3 \times 11) = 623$

$8 \times 38 = 304$ g of $NaBH_4$ produces 623 g of $Ni_{10}B_3$.

1 g of $NaBH_4$ produces $623/304$ g of $Ni_{10}B_3 = 2.05$ g

1.0 kg of $NaBH_4$ produces 2.05 kg of $Ni_{10}B_3$

WHY DOES IT MATTER?

Borazole

Borazole has the formula $B_3N_3H_6$. Its structure is a six-membered ring with alternating nitrogen and boron atoms. It is isoelectronic with benzene as each nitrogen atom has one electron more than carbon and each boron atom has one fewer.

However, unlike benzene, it is quite reactive. This is because the B—N bond is polar ($B^{\delta+} N^{\delta-}$) and can be attacked by polar reagents while the C—C bond is completely non-polar.

THE PERIODIC TABLE

Key

| Relative atomic mass |
| Symbol |
| Name |
| Atomic number |

Period

Group	I	II		III	IV	V	VI	VII	0
1	1.0 **H** Hydrogen 1								4.0 **He** Helium 2
2	6.9 **Li** Lithium 3	9.0 **Be** Beryllium 4		10.8 **B** Boron 5	12.0 **C** Carbon 6	14.0 **N** Nitrogen 7	16.0 **O** Oxygen 8	19.0 **F** Fluorine 9	20.2 **Ne** Neon 10
3	23.0 **Na** Sodium 11	24.3 **Mg** Magnesium 12		27.0 **Al** Aluminium 13	28.1 **Si** Silicon 14	31.0 **P** Phosphorus 15	32.1 **S** Sulphur 16	35.5 **Cl** Chlorine 17	39.9 **Ar** Argon 18
4	39.1 **K** Potassium 19	40.1 **Ca** Calcium 20		69.7 **Ga** Gallium 31	72.6 **Ge** Germanium 32	74.9 **As** Arsenic 33	79.0 **Se** Selenium 34	79.9 **Br** Bromine 35	83.8 **Kr** Krypton 36
5	85.5 **Rb** Rubidium 37	87.6 **Sr** Strontium 38		114.8 **In** Indium 49	118.7 **Sn** Tin 50	121.8 **Sb** Antimony 51	127.6 **Te** Tellurium 52	126.9 **I** Iodine 53	131.3 **Xe** Xenon 54
6	132.9 **Cs** Caesium 55	137.3 **Ba** Barium 56		204.4 **Tl** Thallium 81	207.2 **Pb** Lead 82	209.0 **Bi** Bismuth 83	(210) **Po** Polonium 84	(210) **At** Astatine 85	(222) **Rn** Radon 86
7	(223) **Fr** Francium 87	(226) **Ra** Radium 88							

d-block

III	IV	V	VI	VII						I	II
45.0 **Sc** Scandium 21	47.9 **Ti** Titanium 22	50.9 **V** Vanadium 23	52.0 **Cr** Chromium 24	54.9 **Mn** Manganese 25	55.9 **Fe** Iron 26	58.9 **Co** Cobalt 27	58.7 **Ni** Nickel 28	63.5 **Cu** Copper 29	65.4 **Zn** Zinc 30		
88.9 **Y** Yttrium 39	91.2 **Zr** Zirconium 40	92.9 **Nb** Niobium 41	95.9 **Mo** Molybdenum 42	(99) **Tc** Technetium 43	101.1 **Ru** Ruthenium 44	102.9 **Rh** Rhodium 45	106.4 **Pd** Palladium 46	107.9 **Ag** Silver 47	112.4 **Cd** Cadmium 48		
138.9 **La** * Lanthanum 57	178.5 **Hf** Hafnium 72	181.0 **Ta** Tantalum 73	183.9 **W** Tungsten 74	186.2 **Re** Rhenium 75	190.2 **Os** Osmium 76	192.2 **Ir** Iridium 77	195.1 **Pt** Platinum 78	197.0 **Au** Gold 79	200.6 **Hg** Mercury 80		
(227) **Ac** † Actinium 89	(261) **Db** Dubnium 104	(262) **Jl** Joliotium 105	(263) **Rf** Rutherfordium 106	(262) **Bh** Bohrium 107	(?) **Hn** Hahnium 108	(?) **Mt** Meitnerium 109					

f-block

*Lanthanides

140.1 **Ce** Cerium 58	140.9 **Pr** Praseodymium 59	144.2 **Nd** Neodymium 60	(147) **Pm** Promethium 61	150.4 **Sm** Samarium 62	152.0 **Eu** Europium 63	157.3 **Gd** Gadolinium 64	158.9 **Tb** Terbium 65	162.5 **Dy** Dysprosium 66	164.9 **Ho** Holmium 67	167.3 **Er** Erbium 68	168.9 **Tm** Thulium 69	173.0 **Yb** Ytterbium 70	175.0 **Lu** Lutetium 71

† Actinides

232.0 **Th** Thorium 90	(231) **Pa** Protactinium 91	238.1 **U** Uranium 92	(237) **Np** Neptunium 93	(242) **Pu** Plutonium 94	(243) **Am** Americium 95	(247) **Cm** Curium 96	(245) **Bk** Berkelium 97	(251) **Cf** Californium 98	(254) **Es** Einsteinium 99	(253) **Fm** Fermium 100	(256) **Md** Mendelevium 101	(254) **No** Nobelium 102	(257) **Lr** Lawrencium 103

s-block

p-block